Validation of Cell-Based Assays in the GLP Setting

Validation of Cell-Based Assays in the GLP Setting

A Practical Guide

Editors

Uma Prabhakar, Ph.D. and Marian Kelley

Centocor Research and Development, Inc., Radnor, Pennsylvania, USA

John Wiley & Sons, Ltd

Telephone (+44) 1243 779777

Email (for orders and customer service enquiries): cs-books@wiley.co.uk
Visit our Home Page on www.wiley.com

Other Wiley Editorial Offices

John Wiley & Sons Inc., 111 River Street, Hoboken, NJ 07030, USA

Jossey-Bass, 989 Market Street, San Francisco, CA 94103-1741, USA

Wiley-VCH Verlag GmbH, Boschstr. 12, D-69469 Weinheim, Germany

John Wiley & Sons Australia Ltd, 33 Park Road, Milton, Queensland 4064, Australia

John Wiley & Sons (Asia) Pte Ltd, 2 Clementi Loop #02-01, Jin Xing Distripark, Singapore 129809

John Wiley & Sons Canada Ltd, 6045 Freemont Blvd, Mississauga, Ontario, L5R 4J3

Wiley also publishes its books in a variety of electronic formats. Some content that appears in print may not be available in electronic books.

Library of Congress Cataloging in Publication Data

Validation of cell-based assays in the GLP setting : a practical guide / edited by Uma Prabhakar and Marian Kelley.
p. ; cm.
Includes bibliographical references and index.
ISBN 978-0-470-02876-6 (alk. paper)
1. Pharmaceutical biotechnology—Laboratory Manuals 2. Animal cell biotechnology—bioassay—Validity—Laboratory Manuals. I. Prabhakar, Uma. II. Kelley, Marian.
[DNLM: 1. Biological Assay—methods—Laboratory Manuals. QV 25 V172 2008]
RS380.V35 2008
615′.19—dc22 2007047633

British Library Cataloguing in Publication Data

A catalogue record for this book is available from the British Library

ISBN 978-0-470-02876-6 (H/B)

Typeset in 10.5/12.5pt Times by Integra Software Services Pvt.Ltd, Pondicherry, India
Printed and bound in Great Britain by Antony Rowe Ltd, Chippenham, Wiltshire
We thank QualTek Molecular Laboratories for providing the source of the cover photograph.

The editors dedicate this book to their families for all their support and encouragement.

Contents

List of contributors

Carlos L. Aparicio, Ph.D., Custom BioPharma Solutions, Beckman Coulter, Inc. 11800 SW 147th Avenue, Miami, FL 33196, USA

Jaime Bald, Department of Clinical Pharmacology & Experimental Medicine, Centocor Research and Development, Inc., 145 King of Prussia Road, Radnor, PA 19087, USA

Hector Battifora, MD., QualTek Molecular Laboratories, 334 South Palteron Avenue, Suite 208, Santa Barbara, CA 9311, USA

Steve Bernstein, PhD., QualTek Molecular Laboratories, 334 South Palteron Avenue, Suite 208, Santa Barbara, CA 9311, USA

Wade E. Bolton, Ph.D., Vice President, Custom Bio/Pharma Solutions, Beckman Coulter, Inc., 4300 N. Harbor Blvd., (M/C E-34-E), Fullerton, CA 92835, USA

Josephine H. Cox, Walter Reed Army Institute of Research, US Military HIV-1 Research Program, Suite 200, 13 Taft Court, Rockville, MD 20850, USA

Cuc Davis, Department of Clinical Pharmacology & Experimental Medicine, Centocor Research and Development, Inc., 145 King of Prussia Road, Radnor, PA 19087, USA

Hugh Davis, Ph.D., Clinical Pharmacology & Experimental Medicine, Centocor Research and Development, Inc., 145 King of Prussia Road, Radnor, PA 19087, USA

Patricia D'Souza, Vaccine Clinical Research Branch, Division of AIDS, NIAID, NIH, 6700-B Rockledge Drive – MSC 7628, Bethesda, MD 20892-7628, USA

Guido Ferrari, Department of Experimental Surgery, Duke University Medical Center, P.O. Box 2926, Durham, NC 27710, USA

Amy Fraunfelter, Department of Clinical Pharmacology & Experimental Medicine, Centocor Research and Development, Inc., 145 King of Prussia Road, Radnor, PA 19087, USA

Christina D. Hamm, Cellular Technology Limited and Department of Pathology, Case Western Reserve University, Cleveland, OH 44106, USA

Marian Kelley, Director of Compliance, Clinical Pharmacology & Experimental Medicine, Centocor Research and Development, Inc., 145 King of Prussia Road, Radnor, PA 19087, USA

Nina Thapa Kunwar, Vaccine Clinical Research Branch, Division of AIDS, NIAID, NIH, 6700-B Rockledge Drive – MSC 7628, Bethesda, MD 20892-7628, USA

Paul V. Lehmann, Cellular Technology Limited and Department of Pathology, Case Western Reserve University, Cleveland, OH 44106, USA

Thomas Lohr, Department of Clinical Pharmacology & Experimental Medicine, Centocor Research and Development, Inc., 145 King of Prussia Road, Radnor, PA 19087, USA

Frank Lynch, Ph.D., QualTek Molecular Laboratories, 300 Pheasant Run Newtown, PA 18940, USA

Marielena Mata, Ph.D., Department of Clinical Pharmacology & Experimental Medicine, Centocor Research and Development, Inc. 145 King of Prussia Rd., Radnor, PA 19087, USA

Persymphonie Miller, Clinical Pharmacology & Experimental Medicine, Centocor Research and Development, Inc., 145 King of Prussia Road, Radnor, PA 19087, USA

Jaymala Patel, Department of Clinical Pharmacology & Experimental Medicine, Centocor Research and Development, Inc., 145 King of Prussia Road, Radnor, PA 19087, USA

Charles Pendley, Ph.D., Department of Clinical Pharmacology & Experimental Medicine, Centocor Research and Development, 145 King of Prussia Road, Radnor, PA 19087, USA

Victoria Polonis, Walter Reed Army Institute of Research, US Military HIV-1 Research Program, Suite 200, 13 Taft Court, Rockville, MD 20850, USA

Uma Prabhakar, Ph.D., Department of Clinical Pharmacology & Experimental Medicine, Centocor Research and Development, 145 King of Prussia Road, Radnor, PA 19087, USA

Manoj Rajadhyaksha, Department of Clinical Pharmacology & Experimental Medicine, Centocor Research and Development, 145 King of Prussia Road, Radnor, PA 19087, USA

Manjula Reddy, Department of Clinical Pharmacology & Experimental Medicine, Centocor Research and Development, 145 King of Prussia Road, Radnor, PA 19087, USA

Marcella Sarzotti-Kelsoe, Department of Experimental Surgery, Duke University Medical Center, P.O. Box 2926, Durham, NC 27710, USA

Magdalena Tary-Lehmann, MD, Ph.D., Cellular Technology Ltd, 10515 Carnegie Ave., Cleveland, OH 44106, USA

Thomas Williams, Department of Clinical Pharmacology & Experimental Medicine, Centocor Research and Development, 145 King of Prussia Road, Radnor, PA 19087, USA

Julie Wilkinson, M.S., Beckman Coulter, Inc, Custom BioPharma Solutions, 11800 SW 147th Ave., MC 21-A01, Miami, FL 33196, USA

Jackson Wong, Department of Clinical Pharmacology & Experimental Medicine, Centocor Research and Development, 145 King of Prussia Road, Radnor, PA 19087, USA

Preface

Technology platforms including cell-based assays are used not only for the identification of new drug targets but also for supporting the analysis of clinical samples during clinical development. The former is a discovery research effort and personnel involved in the conduct of the science at this stage must have thorough and extensive knowledge of the technology and science and must exhibit a high degree of integrity in the conduct of the science. Targets identified during this phase impact the commercial and clinical development aspects thereby influencing the drug pipeline for any given pharmaceutical organization. Just as the conduct of clinical trials must strictly adhere to Good Clinical Practices (GCP), so also the analysis of clinical samples in trials must be conducted under very strict surveillance as these efforts directly influence the trial design and eventually affect patient lives. For all diagnostic testing performed on humans in the U.S., excluding clinical trials, Congress passed the Clinical Laboratory Improvement Amendments (CLIA) in 1988 establishing quality standards to ensure the accuracy, reliability and timeliness of patient test results regardless of where the test was performed.

However, a number of the laboratory tests performed on clinical specimens in the present environment include exploratory evaluations aimed at identifying indicators of exposure or susceptibility to drug agents, or at predicting the incidence or outcome of disease. These indicators or "biomarkers" can be soluble or cell-associated and can be measured in whole blood specimens, purifed cell subsets or serum. Following rigorous validations, some of the biomarkers can eventually land up as companion diagnostics or serve to stratify a specific population deemed as being responsive to a specific drug agent/disease indication.

While the standards used for regular laboratory testing can be applied to soluble biomarkers ensuring reliable and reproducible results, the standards for conducting cell-based assays (functional and non-functional) are not well defined by GCP, CLIA or Good Laboratory Practices (GLP). Furthermore, cell-based testing is rather complex as it represents a complete biological system in itself. Therefore establishing a CLIA/GLP-like standard for cell-based testing is not without its challenges and frustrations.

Our first such challenge occurred 5 years ago when we started to develop cell-based assays for evaluating the cellular immune function in patient samples following treatment with immunomodulators. The complexity of developing and optimizing these assays was daunting and confounding in the beginning; nevertheless we undertook the initiative of adapting assays initially meant for discovery work to support clinical trials according to GLP guidelines. The scientists involved in these efforts had prior experience working in a GLP environment. Eventually, the Director of Compliance of the Department of Clinical Pharmacology and Experimental Medicine at Centocor Inc. provided substantial oversight to ensure that best practices were adopted and followed as these assays were being developed. After much arduous and painstaking effort, our laboratory has established assay validations and methodologies for a variety of cell-based assays, and also developed several procedural documents that can serve as a valuable resource for any researcher who is interested in conducting cell-based assay work for clinical trials.

This book contains procedural documents we have developed to describe factors that should be taken into general consideration while setting up cell-based assays, and for the development, optimization and validation of cell assays. A number of actual validations are presented including ELISPOT, flow metric analysis, proliferation and neutralization of immune responses. We also have valuable contributions from several experts in the field who have provided their viewpoints for developing ELISPOT assays, intracellular cytokine assays, immunohistochemistry analysis, and endpoint assays for HIV-1 vaccine trials.

Our incentive to publish this book is solely to provide the professional in the field examples of specific validations for complex cell-based assay platforms and their use in supporting clinical analysis of samples in a GLP setting. The editors do not claim that the methods and procedural documents presented here represent approved regulatory documents. Rather, they reflect best practices that should be followed to ensure consistent and reliable results along with good documentation practices. Our hope is that this book will serve as a living document and as new technology platforms become available, the procedures and practices are updated periodically. Continued efforts to reflect best laboratory practices will ensure the quality data at all times which in the long run will result in quality treatment for patients.

We thank all the scientists in our laboratory for their commitment and painstaking efforts in developing validated assay procedures in this rather complex area and to our internal and external contributors for their valuable inputs.

Uma Prabhakar
Marian Kelley

Introduction

Uma Prabhakar
Centocor Research and Development, Inc., 145 King of Prussia Road, Radnor, PA 19087, USA

Cell-based functional assays have played a major role in biological research and development, starting from target discovery and continuing through pivotal clinical trials for registration of novel drug agents. The core of the assay is the cell composed of hundreds of complex molecules that regulate the pathways necessary for vital cellular functions. By their very nature, cell-based assays are inherently variable and require extra care to achieve consistent performance. They are extremely sensitive to changes in the cell culture medium and to various factors including passage number, temperature, and the surface on which they are grown to name a few.

An early part of the experimental process during drug discovery involves screening a large number of compounds in an ultra high throughput format. It is recognized that the effect of the drug on an organism is complex and involves multiple levels or stages of interaction that cannot be mimicked by using biochemical assays alone. Understanding the complexity at the cellular level, so as to better predict the physiological relevance and impact, requires the use of cell-based assays. Needless to mention, *in vitro* cell-based assays are only an approximation to the *in vivo* physiological setting. Nevertheless, eukaryotic cell cultures are well accepted as the model system of choice to get a first approximation of *in vivo* activity. Advances in assay chemistries and signal detection technologies have allowed miniaturization of cell-based assays, making it convenient to perform a range of experiments, including dose-response etc, during the primary screens.

Cell-based assays are used to assess a variety of cellular aspects including viability, cytotoxicity, apoptosis, signal transduction and metabolic functions. As with any other assay, choice of the cell-based assay is based on the information that has to be measured at the end of the treatment period. During the screening stage of drug discovery, and regardless of the model system chosen, it is important to establish a consistent and reproducible procedure. The number of cells per well, equilibrium period prior to assay (which may affect cellular physiology), maintenance and handling of stock cultures, assay responsiveness to test agents, culture medium, surface to volume ratio, gas exchange, edge effects etc., are some of the factors that have to be kept in mind as these assays are developed. In general, the screening stage is relatively "uncontrolled and undisciplined", since innovation is the key aspect of drug discovery.

While the development and analysis requirements for screening cell-based assays during the drug discovery stage are well defined, the requirements for the assays used to support downstream drug development activities, such as establishing a master cell bank (for biologic drugs) or for evaluating clinical responses to a drug, need to be far more stringent. Furthermore, these assays must also be closely monitored to ensure consistent and robust performance. For establishing master cell banks used to produce biologic drug products, Good Manufacturing Practice (GMP) regulations are established by the Department of Health and Human Services of the Food and Drug Administration (FDA) which require that manufacturers, processors, and packagers of drugs, medical devices, some food, and blood take proactive steps to ensure that their products are safe, pure, and effective. GMP regulations address issues including recordkeeping, personnel qualifications, sanitation, cleanliness, equipment verification, process validation, and complaint handling. Therefore, GMP protects the integrity and quality of the manufactured product intended for human use. Similarly, Good Laboratory Practices (GLP) protects the quality and integrity of the laboratory data used to support a product application. GLP applies when a non-clinical laboratory study (non-clinical animal testing) is intended to support an application for an FDA-regulated product.

However, for cell-based assays used for measuring endpoints in non-primate toxicology studies and in clinical trials, there are no specific guidelines that dictate requirements necessary to qualify such assays. Typically, the Guidelines of the International Conference on Harmonization (ICH) are followed for the validation of different assay parameters including analytical recovery, precision, sensitivity, specificity, selectivity, and robustness. Every effort is made to ensure that "quality" is built in to ensure that the assay is consistent and meets the same specifications time after time.

Our laboratory supports the identification and characterization of pharmacodynamic biomarkers and immunogenicity for our therapeutic biologic drugs. We have developed and validated numerous cell-based assays, in our laboratory, to support biomarker assessments in our clinical trials. Using whole blood specimens, blood products like peripheral blood mononuclear cells (PBMCs), or serum/plasma and tissue biopsies, these assays have been used for a variety of different purposes including, (1) evaluation of the immune status of subjects following treatment with DNA vaccines, (2): evaluation by immunohistochemistry (IHC) changes in the expression of biomarkers, (3) evaluation of the expression of cell surface markers by flow cytometry, (4) characterization of the neutralizing capacity of immune response to our antibody drug-products, and, (5) evaluation of cellular (CD19+ lymphocyte) activation.

To our knowledge, there is no documented guidance available to define the parameters required to establish a qualified cell-based assay in the GLP setting. A subcommittee of the AAPS Ligand Binding Focus Group (LBABFG) published their recommendations (DeSilva et al, 2003) for the development, validation and implementation of ligand binding assays (LBAs) that are intended to support pharmacokinetic and toxicokinetic assessments of macromolecules. The recommendations in this publication are based on bioanalytical best practices and statistical thinking for development and validation of LBAs. Another recent publication (Gupta et al, 2007) provides recommendations on the development, optimization and qualification of cell-based assays for assessing the neutralizing capacity of anti-drug product antibodies by using a fixed concentration of drug in the neutralizing antibody assay (Nab). The recommendations are based on the authors' experience and reflect scientific concepts to assist assay developers form a rationale for the development of their specific assay.

Using the LBA recommendations mentioned previously, the Guidelines of the ICH, the white paper (DeSilva et al, 2003) and the Nab assay recommendations (Gupta et al, 2007) as our reference points, we set out to define parameters required to make cell-based assays compliant with GLP so that the data generated could support an application for an FDA-regulated drug product. We developed procedural documents for the development of bioassays or cell-based assays where cell lines are used to measure the quantity and or functional activity of analytes present in a biological matrix. Procedural documents were also prepared defining criteria for validating immunoassays used to evaluate cellular responses using peripheral blood mononuclear cells obtained from subjects. Using these procedural documents and guidelines several cell-based assays were successfully developed and validated and are being used in support of clinical trials. The validation reports contain detailed information on how the experiments were planned and conducted, the process for reporting the results and all the documentation procedures that have to be in place for the data generated. A number of investigators who adopt these procedures in their laboratory, for cell-based assays, provide their perspective of development and validation of such assays.

As new technology platforms become available for cell-based assays, the procedural documents will need to be modified accordingly. At the same time, it must also be recognized that not all cell-based assays may lend themselves to GLP nor do they necessarily have to be conducted under such compliance since they are purely exploratory. This is particularly true with some of the 'omic's technologies, using cell derived lysates, which are currently being used for target identification and biomarker panning purposes.

In the following chapters of this book, a practical guide for conducting a variety of cell-based assays is available for a reader not familiar with GLP, or who wants to set up cell-based assays in their laboratory, or assess contract vendors who provide such assay services. This guide does not reflect any FDA regulations or guidances and is based on the authors' personal experiences in the use and conduct of cell-based assays.

References

DeSilva B, Smith W, Weiner R, Kelley M, Smolec J, Lee B, Khan M, Tacey R, Hill H and Celniker A (2003). Recommendations for the bioanalytical method validation of ligand-binding assays to support pharmacokinetic assessments of macromolecules. *Pharm Res*, **11,** 1885–1900.

Code of Federal Regulations part VI. Department of Health and Human Services Food and Drug Administration.

Gupta S, Indelicato SR, Jethwa V, Kawabata T, Kelley M, Mire-Sluis AR, Richards SM, Rup B, Shores E, Swanson SJ et al (2007). Recommendations for the design, optimization, and qualification of cell-based assays used for the detection of neutralizing antibody responses elicited to biological therapeutics. *J Immunol Methods*, **321**, 1–18.

1 Considerations while setting up cell-based assays

Marian Kelley
Clinical Pharmacology & Experimental Medicine, 145 King of Prussia Road, Radnor, PA 19087, USA

1.1 Introduction

Cell based assays appear to be increasing in number and importance within the Pharmaceutical Development arena. There are innovative new platforms available, along with historical and established methods using cells as an integral part of the assay design. The data generated by these assays are being used to support such diverse endeavors as the characterization of an immune response in support of pharmacokinetics or phase IV safety, measures of cell activation, proliferation, and death, cell surface marker expression, and confirmation of useful biomarkers thereby contributing to decision-making in the Drug Developmental process.

Because both the assays themselves and the intended use of the data are so diverse it is difficult to standardize a single validation strategy. The stage of the development process, sample and data types, and how the data will be used, all influence what elements will be included in assay development and the validation plan. Generally speaking it is prudent to

Validation of Cell-Based Assays in the GLP Setting: A Practical Guide Editors, Uma Prabhakar and Marian Kelley

require the data used as the basis for scientific or clinical decision making to be both accurate and reliable. To most effectively accomplish this it is recommended to incorporate a general Good Laboratory Practice (GLP) format for performing a defined assay development program, which is used to generate a validation plan incorportating, *a priori* acceptance criteria.

This chapter will discuss the development and validation of cell-based assays including the lead in to development, basics that should be accomplished during development, execution of the validation plan and the final report.

1.2 Lead in to assay development

(A) Cells: The major requirement to develop a cell-based assay is the cells themselves. Cell function assays such as those investigating cytokine secretion, apoptosis and cell-surface marker modifications depend on a reliable source of cells, whether fresh or cryo-preserved, to make accurate and defensible conclusions. Cellular histology platforms also require a systematic and well-defined procedure for collection and preparation to ensure the consistent and reliable source of cells. When the source of cells is patient samples it is imperative that well thought out processes for collection, shipment and storage are implemented to ensure accurate and reliable data.

In some cases, such as proliferation and neutralizing antibody (NAB) assays, a cell line is used as the source of the assay read-out. Without a secure source of this cell line that can be expected to provide a consistent assay reagent the entire development and validation process is compromised.

A description of the cell line is essential and should include how it was developed, media and growth conditions, storage and recovery. References are useful if available. A full description of the cell banking process is helpful. When using cell-lines for a cell-based assay special care must be taken to preserve the cells by creating a Master Bank. Cell Banking is performed to preserve the characteristics of the cell line to be used. It is recommended that the cell-banking program be implemented as early as possible in the life cycle of assay development. Often it is not known how many passages a cell line can withstand before drift occurs. Cell banking also offers insurance that the cell line can be re-established in the event of a catastrophe like microbial contamination, cross contamination with other cell lines, or loss of desired characteristics.

As soon as a cell line is introduced into the lab an initial Master Bank should be frozen. The number of ampoules will be dependent on how quickly the cells multiply but at least 3–5 ampoules should be frozen within the first week. Initial evaluation of the cells should be as complete as possible, but at least examine sterility (mycoplasma, fungi, etc) growing conditions, viability, and the ability to be frozen and recovered from liquid nitrogen. Once convinced of the cell line's integrity, the main Master Bank should be prepared by thawing an ampoule from the initial Bank. Since it is preferable to expand these cells to a high concentration with as few passages as possible, while maintaining high viability, it may be advisable to thaw several of the ampoules, if the inventory of the initial Master Bank allows it. The number of vials contained in the Master Bank is dependent of the life expectancy of the assay. It is always prudent to bank more than the expected requirement, even if the Master Bank must consist of several different and increasing passage numbers due to the slow growth of cells. Once established, the Master Bank is used as the supply for the Working Stock. Early in the process while the working stock is in culture the number of passages should be monitored closely and tested at intervals to determine that its integrity is being maintained. Well before the Master Bank is depleted a sequential Master Bank should be prepared, if necessary.

Tests critical to the determination of the continued integrity of the cell lines' required characteristics should be conducted thoughtout the cells' expansion to assess the optimum permissible passage number.

If the cells once thawed lose viability some rescue methods may be employed. Dead cells can be removed by centrifugation or other method. Cells can be nursed to higher viability by expanding in smaller culture volumes/ culture plates. Higher concentrations of sera, if used, or other growth supplements, may encourage growth. Be aware that such rescue methods could encourage the growth of a variant cell line and further re-characterization would have to be performed.

(B) Assay format: The type of assay to be used will define the development process and validation needed. The data generated by the assay may be quantitative and consist of a continuous numerical value, such as data reported from a regression of a standard curve. Proliferation assays frequently are reported based on this format. Other qualitative formats allow for a discrete or descriptive, numeric-reporting format, where the data is spaced across the axis or used a descriptive, non-numeric term (e.g., high or low; yes or no). Of course, intrinsic to the assay format is the sensitivity requirement. This must be determined at the initiation of development, based on the intended use to confirm that the platform selected and data reporting will afford the sensitivity to meet the needs of the study.

(C) Critical Reagents: It is important to identify which reagents are critical to the assay method so that their qualification, sourcing, and lot-to-lot acceptance criteria can be established up-front. Additional assessments conducted during this phase include read-out signal (color intensity, MTT etc) incubation times, reagent concentrations etc.

1.3 Assay development

The output of the development lead in described above is a high level plan forward. The development phase is typically the most intensive and results in a defined method that enters the validation stage.

Cell-based assays differ from ligand-binding methods and can be characterized based on the type of assay format or platform. They also differ significantly from each other since they may consist of a single "layer" or be multi-layered. The simplest example is a one-layered cell-based assay. This type describes the immunohistological slide platform, or the cells line with single stimulus, e.g., a cytokine's effect on a cell-line, which elicits an expected response. An example of a one-layered cell-based assay uses the agonist cytokine where a dependent cell line proliferates in a dose-dependent manner to the addition of increasing amounts of the cytokine.

Some assays developed to detect neutralizing antibody build upon the one-layered assay by adding an inhibitory facet to the proliferation assay mentioned above. An example of a two-layered cell-based assay is the assay to detect antibodies to a cytokine therapeutic. In this case the method would include the cell line, its optimized stimulatory element followed by a serial dilution of an expected inhibitory element such as patient sera or spiked quality control samples containing antibody to the therapeutic protein.

When antagonists are being developed as a biologic therapeutic, frequently the complementary neutralizing cell-based assay must be developed and is composed of three layers. This is the case for some monoclonal antibody (MAB) therapeutics since the action of the antagonist monoclonal is to inhibit the action of a stimulus (the MAB- related agonist) on the cells. Assays developed to detect neutralizing antibodies to the MAB would add an additional layer to the basic cell-based assay. The "normal" process of a responsive cell type responding to the target of the MAB would be inhibited by the addition of an optimized amount of therapeutic drug. The cellular response to the target is salvaged by the addition of samples containing varying concentrations of antibodies to the monoclonal therapeutic (see figure one: three-layered assay).

Assays using cells that constitutively produce a cytokine, for example, may be referred to as four-layer assays since the basal concentration of cytokine (1) is another parameter that would need to be monitored during validation and sample analysis. In this instance the basal concentration of cytokine may be enhanced (2) with the addition of a specific cytokine and the therapeutic drug would reverse (3) that increase. Detection of neutralizing antibodies (4) to the therapeutic drug comprises the fourth layer (see figure two: four-layered assay). Specific monitoring for each step is necessary to assure the consistent behavior of the method.

The "formula" for optimizing a cell-based assay method is guided by the number of layers attributed to the method. All the layers leading up to the final "read-out" must themselves be optimized sequentially and in a way that permits the final "read-out" to be useful. The final "read-out" must be a dose-dependent response (quantitative or qualitative) that is attributable to and can characterize the test in question. The requirements for the optimization of the layers are likely similar to each other. The cell number, concentration of agonistic therapeutic (layer 1), then concentration of inhibitory factor (layers 2 or 4), or concentrations of antagonistic therapeutic (layer 3) must be tested in a dose-dependent manner to select the optimal dose (concentration) to be used in the final format. It is insightful to understand that the more layers an assay contains the more complicated the selection of the optimal concentration for each layer. The concentration that produces the highest response is frequently not the best choice. For instance, in cases where sensitive neutralizing antibody detection is needed, the aim is to detect low concentrations of antibody. Adding in very high concentrations of drug to be neutralized skews the assay to require a high antibody response rate. To be able to detect low antibody response rates, the method developer will need to balance the added drug to be neutralized by the lower apparent antibody present together with an acceptable response range.

Other parameters that are optimized during development but are independent of the layering aspect of the assay method include the cells themselves, i.e., cell passage, viability, sensitivity in the presence of subject sera, response variability etc.

All the elements up to now have been performed in development. The final assay method now becomes the focus of the validation stage. By compartmentalizing the development in this fashion, the validation experiments may become focused on documenting a reliable, robust and reproducible assay.

1.4 Sample handling

Special attention must be paid to how the samples targeted for analysis in cell-based assays are collected, processed, stored, shipped, or frozen. Since each of these conditions is dependent on the platform to be used, the specifics of sample handling are best presented in the context of the particular assays described in this manual. A description of the investigation into appropriate sample conditions are documented the validation report.

1.5 Validation plan and conduct

Validation is typically preceded by a validation plan, which summarizes *a priori*, the performance parameters to be tested. The extent of the validation and the acceptance criteria are dependent on several factors, among them, the needs of the study, the nature of the methodology, and the observed variability (Lee et al, 2006). Generally, the stringency of the validation parameters should correlate to the drug development stage where the assay is to be used. Less rigor would be expected for assay supporting Drug Discovery or early development. In most of the cases in this manual the focus is on clinical samples support of late stage clinical studies. Therefore, in following the stage-appropriate validation a more inclusive validation would be expected.

Once the validation is initiated experiments are expected to proceed uninterrupted and the experimentation documentation should reflect that. Analysts must be alert to cases when the assay method fails. One failure is likely not a cause for concern; however, there should be a plan for when failures do become a cause for concern and an investigation into the cause is required. At this point it should be clear in the documentation that the analyst has moved out of validation and back into development or failure investigation. Once the issue is resolved, a determination is made whether the resolution had a minor or major impact on the validation. If minor, the documentation should reflect a return to the on-going validation. If major, note that the original validation failed and a new validation must be implemented. After completion of the described experiments a validation report is required that captures the performance of the assay and any deviations from the described assay method or validation plan.

The validation plan may include:

- Introduction including purpose of the assay
- Background information
- Description of the assay and critical reagents
- Description of validation experiments
- Target criteria for the validation parameters to be included
- Positive and negative controls for each layer of the assay are needed to monitor the assay robustness
- Analysts conducting the validation
- Data handling technique
- Notebook and raw data references for assay development
- Archival location
- Management approval

Validation experiments

Controls

- Positive and negative controls for the cell-based assay method are used to monitor the robustness of the underlying assay and accept a run.

- Validation controls are used to assess the parameters of the assay to support the claim of validated method. The validation controls should reflect the intended samples, typically in human or animal serum.
    - Intra- and inter-assay precision including between runs, days and operators
        - Validation controls prepared using unique donors are assayed multiple times in a run, and over several days, conducted by several analysts to assess the precision of the replicate controls. The positive and negative controls used to monitor each layer of a cell-based assay may also be assessed to document the overall precision of the assay.
    - Assay cut-off to determine sensitivity or the difference between an positive and a negative sample
        - It is recommended employing as many unique donors (e.g., animal, normal human or target disease populations) as possible and in several assays to determine the appropriate cut-off. Adding two standard deviations to the mean read-out provides a false positive rate of about 5%, which ensures an acceptably sensitive assay.
        - Sensitivity may also be determined empirically by spiking quality control at a high concentration and titrating in several assays. The sensitivity is the lowest titer (or concentration) of the quality control (QC) that produces a value with acceptable precision. This experiment may also establish dilutional linearity of the sample.
    - Assay range and limits of quantification, if relevant, including the lower and upper limits (LLOQ and ULOQ)
        - When an assay is quantifiable, the standard curve range and upper and lower limits using spiked controls are assessed. Every run employed for the validation that includes the standard curve and independently prepared quality control samples should be compiled in two tables to document the overall performance of the curve and the controls during the validation.
    - Specificity and Selectivity
        - These parameters are closely related and are assessed to verify that the assay is specific for the intended use (will not tag a closely related but unintended target) and can preferentially select the intended target from a complicated milieu. While assay cut-off experiments are conducted in unspiked target matrix, these experiments employ multiple spiked matrices.

  - The impact of drug interferences can also be assessed during the investigation into specificity.
  - Specificity of the cell line, if applicable, is a parameter unique to the cell-based assay platform. Especially when the method makes claims of responding specifically to a cytokine or other stimulus this claim must be supported by testing the cells in the presence of factors found in a relevant matrix.

- Robustness

  - To understand the inherent reproducibility of the method, the impact of typical changes and varied conditions that can occur during sample analysis is assessed. The conditions tested depend on the assay format, and can include such parameters as incubation times and temperatures, cryopreservation and histology techniques, matrices etc.

- Stability

  - Cell-based assays, as described in this manual, have very specific requirements depending on the platform used. In all cases some investigation into the stability of the target in the milieu selected, (e.g., whole blood, peripheral blood mononuclear cells, tissue samples for histology, serum etc) must be conducted to assure the validity of the data reported.
  - Some references are also made to stability of response. To understand and anticipate the variability expected, some investigation should be conducted on the stability of the target response, e.g., cell surface expression on tissues on fresh, shipped, frozen/thawed and preserved samples, etc.

1.6 Validation report

Once completed and the experiments conducted to establish the validity of the assay are found acceptable it is necessary to write a report to document the assay validation. As a suggestion, a validation report typically contains an introduction and history of the assay to date. Also important to include are dates of the conduct of the validation, references to the analysts involved and the raw data notebooks to support assay reconstruction and where they are archived, a description of the experimental investigation and tables supporting the validation, any deviations made to the original validation plan. The report should be signed by the author and management and centrally archived for easy retrieval.

1.7 Conclusion

When data generated using cell-based assays are used as the basis for scientific and clinical decision making, investigators must apply and rigorously monitor appropriate controls for all "layers" of the assay to ensure its continued validity. Additionally, assay validation is a dynamic process. It is expected that questions will arise over the time the assay is being employed, for instance, as new disease states are being investigated. By applying the concepts of GLPs cited here the resulting assay should be a well-developed and well-documented method validation.

Reference

Lee JW, Devanarayan V, Barrett YC, Weiner R et al (2006). Fit-for-purpose method development and validation for successful biomarker measurement. *Pharm Res*, **23**(2): 312–328.

2 Development, optimization and validation of cell-based assays – 1

Marielena Mata and Thomas Lohr
Centocor Research and Development, Inc., Radnor, PA 19087, USA

2.1 Introduction

The increase in biomedical research in recent years has not translated into the desired outcome, i.e., increased new products in the market, largely due to uncertain results coming from current test procedures (FDA, 2006). New methods are required to better understand the biological effects our developmental compounds/agents have on our subjects. Similarly, we also need to understand the effect our subjects have on the new drugs. Cell-based bioassays have the potential to evaluate both the effect of new drugs on

* Address correspondence to: Marielena Mata, Centocor R&D, Inc. 145 King of Prussia Rd., Radnor, PA 19087, USA. Phone: (610) 651-6037. Fax: (610) 651-6262.

Validation of Cell-Based Assays in the GLP Setting: A Practical Guide Editors, Uma Prabhakar and Marian Kelley

patients, in the form of functional pharmacodynamic assays and biomarkers, and the effect of patients' responses on new drugs, for example, neutralizing immune response assays. In order to develop functional bioassays, a number of approaches must be considered. One approach to evaluate a product is to measure the effect of the product on subject cell samples. For example, DNA vaccines can be evaluated by monitoring the patient's T cell population with such methods as antigen-specific T cell proliferation assays, Enzyme-linked Immunospot (ELISPOT) assays or flow cytometry (Keilholz et al, 2002). The enumeration of circulating tumor cells (CTCs) is currently being evaluated as a surrogate marker for response to cancer therapeutics (Cristofanilli et al, 2004, Cristofanilli and Mendelsohn, 2006) and T cell activation is evaluated as a measure of immune function following transplantation immunosuppressive regimens (Kowalski et al, 2003). Alternatively, a more traditional approach is to use cell lines that respond to the analyte in question in a given biological matrix. Such an approach can be applied to measure bioactive levels of hormones and cytokines, and to evaluate the neutralizing effect of antibodies against biological drugs. An important aspect of implementing these methods in the clinic and in the overall development plan of any drug, is the establishment of a validated assay in compliance with guidelines established by the FDA and other regulatory agencies (FDA, 1994). These guidelines are not specific for the development and validation of cell-based assays, as they are targeted towards traditional bioanalytical methods.

In this chapter, we present a procedural document to guide the validation of bioassays or cell-based assays, in support of both non-clinical and clinical studies, where cell lines are used to measure the quantity and/or functional activity of analytes present in a given biological matrix.

2.2 Definitions

The following definitions have been used in this chapter:

Control Analyte: A substance used to demonstrate assay performance or to represent a (real/mock) sample

Sample Analyte: A clinical or non-clinical tissue specimen generated during the study of a test article

Quantitative: A result that is reported in continuous units due to comparison to a reference standard

Semi-quantitative: A result that is reported in relative units due to an absence of a reference standard

Qualitative: A result that is reported as positive or negative due to an absence of a reference standard

Positive Control: A control that establishes that the assay worked by providing an expected positive result

Negative Control: A control used to establish the baseline response of the assay

Mean: the sum of all the observations divided by the number of observations

Standard Deviation (SD): The square root of the variance as defined by the following formula:

$$\mathrm{SD} = \surd[\sum(x - \mathrm{mean})^2/(N-1)]$$

where x is an individual value and N is the sample size.

Coefficient of Variation (CV): Defined by the following formula:

$$\mathrm{CV}=\mathrm{Standard\ Deviation/Mean}\times 100$$

2.3 Assay development and optimization

During the development of a new drug, a number of questions arise that can best be answered with a functional bioassay: Is the drug target functional? Is the drug being neutralized? Is the drug having the desired pharmacodynamic effect? Based on these questions, assay formats need to be evaluated prior to development. There may be several approaches taken and each approach should be considered carefully in order to choose an assay format that can be carried forward. Some criteria to be evaluated when choosing an assay should include sensitivity of the assay, throughput and complexity of the assay (e.g. number of steps and reagents). Once the assay format has been defined, the development and optimization of the assay should include testing several assay parameters that result in responses that reflect good signal to noise ratio.

The first step in development of an assay requiring a cell-line readout is to select an appropriate cell line for such assay. Factors to consider include availability of the cell line, stability of response based on the current literature and true dependence on the analyte. Once the cell line has been identified, culture conditions should be optimized. Conditions such as passage number, cell number and media conditions affect cell-line responses to the analyte of interest. Therefore cell culture conditions that are appropriate and optimal for the factor to be measured should be selected from evaluation of studies of such conditions. The appropriate length of stimulation that is feasible (taking into consideration the sample and/or cell-line viability) and one that excludes non-specific effects should be determined. The appropriate temperature of stimulation (usually 37 °C) and suitable equipment that best maintains this temperature, such as, 37 °C water bath, dry heating device, dry heat 37 °C CO_2 incubator should be evaluated. The specificity of the response elicited by the stimulation reagents should be assessed using an irrelevant control or a blocking agent. Specificity should also be evaluated in the presence of other interfering factors such as other cytokines and therapeutics that are likely to be present in the test samples.

In bioassays, the cell lines used to measure a response are a critical component and one subject to the most variability. Given the variability inherent with a biological/cell-based assay, acceptance criteria for many of the parameters tested have to be established during optimization but remain more flexible (see section on Acceptance Criteria). Nonetheless, certain parameters should be evaluated to ensure reproducibility. While evaluation of these parameters is not part of the validation, the conclusion obtained from the data may be reported in the validation as a requirement for the assay. The most critical parameters to be evaluated are passage number and cell density, incubation conditions and matrix effects on the cell line. As the guidelines in this chapter are applicable to a number of cell types and assay platforms, the supervisor and operator should be responsible to determine how to best evaluate these parameters based on the requirements of the assay.

After optimization and before the beginning of the validation testing, cells should be cultured, expanded and cryopreserved in liquid nitrogen for further use in a working cell bank. This cell bank should include cells cryopreserved at different passages previously shown to give optimal results. Viability of the cells should be determined for at least the first and last passage of the cryopreserved cells. In addition, to determine the effect of the cell passage number on the assay, cells from at least two different passage numbers (low and high) should be thawed and tested in the assay for comparable results.

To determine the effect of cell density on the performance of the assay, two different evaluations are performed. The first one is to evaluate the cell density of the cell culture before assay set up. During this evaluation, both cell density and viability of the cells should be considered when looking at the assay performance. The second one is to evaluate the number of cells in the well during the assay using a range of cell densities.

Optimization of the incubation times required for the assay, include starvation time, incubation with biological matrix or with other reagents. Incubation times should be varied to cover a range of times within which the assay performs well in the range of the desired parameters.

The effects of the sample matrix (typically serum or plasma) should be evaluated to ensure that the matrix does not have a direct effect on the cell line with regards to viability or response to analyte. To evaluate the effect of the sample matrix, different dilutions of the sample matrix may be tested for optimal results.

Finally, other parameters that may need to be evaluated and optimized for the assay include standard curve, controls and stability of responses. When developing a quantitative assay, standard curves should be optimized to maximize the number of points that fall on the linear part of the curve while providing enough anchor points on both ends of the curve. Controls should include at least one positive and one negative. The response from a negative control may be subtracted from all test and control results, i.e., used as a blank. The stability of responses and other functional parameters should be examined in samples that mimic clinical study sample receipt and compared to samples tested immediately after blood draw. Samples drawn at different times of the day should be tested to assess diurnal variation.

2.4 Assay validation

Once an assay has been optimized, a validation plan including the established acceptance criteria should be prepared. The validation plan should include the following sections when relevant: Controls, Standard Curve, Stability, Specificity and Precision (Table 2.1)(FDA, 2001).

Preparation of controls

Controls should include at least one positive and one negative control. A diluent control may also be included. The diluent may be subtracted from all test and control results, i.e., used as a blank.

A positive control is a known amount of purified analyte of interest added to the matrix or samples. At least two levels of positive controls should be used. The high level positive control would have a result near the middle of the response range. The low level positive control would have a response roughly 2 to 3 times that of the assay cut-off or the negative control.

A negative control is commonly a pool or individual normal human or animal sera targeted patient population, or an animal representing a specific disease model.

Controls may also be used to characterize assay sensitivity, specificity, and ruggedness, as discussed in other sections of this chapter.

Standard curve performance

In quantitative assays, a standard curve performance is evaluated by comparing the mean-back calculated concentration value of each standard (reference) sample with its theoretical (nominal) value. The acceptance

Table 2.1 Summary of main components of a validation report for cell-based assays using cell lines

Component	Parameter	Conditions
Preparation of controls	Positive control Negative control	
Standard curve performance	LOD LLOQ ULOQ Assay cut-off	
Stability	Sample and control	Freeze/thaw (at least three cycles) 4 °C storage (up to 28 days) –70 °C storage (up to 28 days) Liquid nitrogen storage (up to 28 days) Stability of diluted samples
	Other critical reagents	Freeze/thaw (at least three cycles) 4 °C storage (up to 28 days) –70 °C storage (up to 28 days) Liquid nitrogen storage (up to 28 days) Stability of diluted reagents
Specificity	Presence of an interfering substance	
	Drug interference Populations	
	Matrix effects on assay	Serum versus plasma Animal serum versus human serum Normal serum donors versus disease (patient) samples Recovery of an experimental spike
Precision	Inter-subject Intra-assay Inter-assay Inter-operator	

criterion is that mean result of each standard sample should be $100 \pm 30\%$ that of the nominal value.

In addition, sensitivity of the assay, defined as the amount of analyte required to produce a significant change in signal versus that obtained in the absence of analyte should be established using the following parameters, Limit of Detection (LOD), Lower Limit of Quantification (LLOQ) and Upper Limit of Quantification (ULOQ) and cut-off values depending on the quantitative vs. qualitative nature of the assay.

The LOD is derived from multiple tests of the assay diluent control mimicking the sample (blank). The calculation of LOD will be dependent on the assay platform used. The supervisor and operator will identify the best way to determine the LOD based on the assay platform. LLOQ and ULOQ are determined by using spiked samples. The lowest and highest concentrations at which the result is still within 30% of the nominal value are established as the LLOQ and ULOQ, respectively. In some instances, an assay cut-off may be established by running multiple samples from naïve individuals of the species that will be used in non-clinical and clinical trials. Normal donor serum/plasma samples may be used if it is not practical to obtain specimens from a specific disease group. It is recommended to use at least five samples for non-clinical and clinical assay validations but some may suggest up to 100 samples if available. In qualitative assays, the cut-off value may be used to differentiate positive and negative results

Stability

The stabilities characterized should include the control and/or sample analyte and other critical reagents. Stability of the cell line used to measure response is evaluated during the optimization of the method. Optimal parameters should be clearly delineated in the validation report.

Sample and control stability

Testing the controls at consecutive time points during a regimen of storage or handling assesses the stability of controls. It may be necessary to use control samples in order to validate sample stability since actual test samples may not exist at the time of validation. It is particularly important to maintain the consistency of positive controls, which should therefore be banked as frozen aliquots. Each time a new control aliquot is thawed, it should be properly dated (with the date thawed) and used within any expiration date determined through stability testing.

Stability of samples and controls should be assessed by comparing freshly prepared controls and samples with those that have been stored under conditions that might be used for non-clinical and clinical trial samples. Common storage conditions evaluated may include:

Freeze/thaw (at least three cycles)
4 °C storage (up to 28 days)
–70 °C storage (up to 28 days)
Liquid nitrogen storage (up to 28 days)

Stability of diluted samples

Samples in a study refer to real clinical or non-clinical specimens, which might not exist during validation, or may be otherwise unavailable for validation testing. If a positive sample is not available, a mock positive sample might be made from the positive control(s). If the positive control(s) is recoverable in a negative sample matrix, then the positive sample(s) can be created by spiking the positive control(s) into individual normal matrix from the species that might be used for the study. Multiple positive samples at different concentrations are preferred. When a positive control cannot be recovered from the sample matrix, the positive control prepared in diluent may serve as a mock sample. Regardless of the ability to recover a positive spike, stability of the positive control should be characterized if the stability of this control may be an issue. The mean recovery results of stored versus fresh positive control and sample are calculated. Only the mean recovery of positive samples needs to be calculated. The acceptance criteria for the positive control and sample are that they generate positive results, and have a mean recovery of $100 \pm 30\%$ following storage.

The negative control may serve as a negative sample in the stability study. Only one negative sample is required for the stability study. The acceptance criterion for the negative control/sample is that it generates a negative result.

In a freeze/thaw study, multiple aliquots of each sample are prepared, the exact number depending on the number of time points to be measured. Prepared samples are stored frozen (preferably at –70 °C) for at least 24 hours, then thawed at 4–37 °C. When completely thawed, the samples should be refrozen to the original temperature for at least 12 hours before repeating the thawing process. The prepared aliquots of each sample go through an increasing number of freeze/thaw cycles. All samples should be thawed at the same time for the final freeze/thaw cycle, and tested against freshly prepared samples (Day 0), which are prepared in the same way as the frozen samples.

Stability of samples and controls stored at 4 °C should be established. Samples or controls examined in this study should be thawed and then stored at 4 °C for up to 28 days. A minimum of two time points (Day 0 and Day 28) are needed, and intermediate time points such as day 14 are recommended. Multiple aliquots of each sample/control are prepared, the exact number depending on the number of time points to be measured. Samples/controls that are normally stored in a frozen state should be frozen (preferably at –70 °C) for at least 12 hours prior to thawing for this test. The first set of aliquots to be thawed and held at 4 °C is used for the 28-day stability time point. A set thawed and stored at 4 °C for 14 days can be included to provide an intermediate time point. The last set of aliquots, thawed 28 days after the first set is the freshly thawed comparator sample/control (zero days of storage at 4 °C) to which the stored aliquots are referenced.

Samples and controls are generally assumed to be stable as long as they are stored frozen at –70 °C or in liquid nitrogen. Since fresh samples are not always available during validation, a mock sample (positive control in negative sample matrix) should be used to compare fresh and frozen stored time points. The mock sample(s) in this study should be divided into two or more aliquots and stored up to 28 days at –70 °C. Additional time points, such as an intermediate analysis on Day 14, are recommended. On the final day of storage, mock samples are thawed and tested against freshly prepared mock samples (zero days of storage).

Non-clinical samples are frequently diluted for analysis in a bioassay. It may be convenient to dilute samples the day prior to analysis and store them overnight at 4 °C if they are stable under these conditions. To test the stability of diluted samples, positive and negative samples/mock samples are diluted to the final dilution used for analysis, and stored at 4 °C overnight. The stored samples are analyzed side-by-side with freshly diluted samples and the results are compared.

Stability of other critical reagents

Stability is not an issue for reagents stored according to the manufacturer's recommendations or the conditions provided by the labs that generated them if the provider has properly evaluated these reagents. Novel reagents prepared or synthesized by the testing lab, such as antigen conjugates and coated plates, may require stability testing to monitor if assay performance changes over time. This can be performed using similar protocols described for sample stability above. If reagent stability is unknown or found to be an issue, it is recommended that critical reagents be frozen in small aliquots that can be discarded a short time after thawing.

Specificity

Assay specificity is the ability to measure the analyte unequivocally in the presence of other components, either exogenous or endogenous. Depending upon the assay format, interference can be caused by matrix components that are present or over expressed in certain relevant disease populations, or substances that bind competitively with analyte.

To evaluate specificity, the test article or other potential interfering substances, can be spiked into positive and negative samples/controls. Positive controls or mock positive samples and negative samples/controls should be prepared with biologically relevant concentration(s) of a potential interfering substance. The samples/controls with and without the interfering substance are then tested and compared against each other. The acceptance criterion is that mean result from the sample/control should be $100 \pm 30\%$ that of the reference value, in which case there is no substantial interference. Otherwise, it should indicate that the substance could interfere with the detection of analyte.

Test article, such as pharmaceutically administered immunoglobulin (drug), is often present in non-clinical and clinical study specimens and can compete for analyte binding when the analyte is the target of the drug administered. Although this competition represents specific binding to the analyte, the observed effect is an apparent reduction of analyte quantity or function. Preparing positive samples/controls with varying amounts of experimentally added drug can mimic the presence of test article in a sample. Testing such samples/controls can provide information about the degree of interference that may occur. A set concentration of the positive control(s) or other available positive samples may be tested relative to a series of drug dilutions.

Interfering substances may be prevalent in some populations, although it may not be possible to identify the interfering substance. To examine specificity as related to a specific population, defined either by a disease state or a biomarker, mock positive samples are prepared by adding analyte into a biological matrix from the reference population, such as pooled normal human serum, and multiple serum/plasma samples from naïve patients from the intended population. The number of samples used in this study depends on availability, but it is recommended that at least three individuals be tested. Recovery of the positive signal is evaluated by comparing the mean results of the patient and reference populations spiked with the same amount of analyte. The acceptance criterion is that the mean result from the patient population should be $100 \pm 30\%$ that of the reference. Otherwise, the validation would indicate some degree of interference.

Matrix effects on assay

In addition to the effect that the matrix may have on the cell line, the matrix can affect the assay itself due to interfering substances that react with the analyte or critical reagents involved in the detection system. In addition, an assay may be developed to fulfill several purposes, i.e. both non-clinical and clinical studies, different disease indications or different sample types. In each case, an assessment of the matrix effects on the assay should be completed. Some of the common assessments for matrix effects include: serum versus plasma, animal versus human, and healthy normal serum donors versus disease (patient) samples.

The effects of the sample matrix (typically serum or plasma but others may be included) should be evaluated if more than one sample matrix will be tested. Matching samples (such as serum and plasma) from the same donor should be compared. Mock positive samples may be prepared by spiking the positive control analyte(s) into normal donor serum/plasma samples. Normal donor serum/plasma can also be used as the negative samples. It is recommended to compare matching matrices samples from five to ten different individuals, including at least two positive donors if available. It is likely that positive samples will need to be represented by mock positive samples. The acceptance criterion is that the matrix should not alter the assay outcome, i.e., this test passes if positive samples are positive and negative samples are negative in both matrices compared.

Matrix differences between animals and humans should be compared if the assay is intended to be used on both sample types. If appropriate samples are available, it is recommended to compare five to ten specimens of each species (animal and human) including at least two positive donors if available. It is likely that positive samples will need to be represented by mock positive samples. The acceptance criterion is that the matrix should not alter the assay outcome, i.e., this test passes if positive samples are positive and negative samples are negative in both matrices compared.

Non-clinical studies typically examine healthy animals, but may include disease models. Clinical studies may encompass normal volunteers and patients. The specific models and populations may not be identified at the time of assay validation. If the assay is expected to test specific populations, and if specimens are available, then normal/healthy and patient/disease samples can be compared. This test should include five to ten samples of each type, including at least two positive donor samples if available. It is likely that positive samples will need to be represented by mock positive samples. The acceptance criterion is that the matrix should not alter the assay outcome, i.e., this test passes if positive samples are positive and negative samples are negative in both matrices compared.

Precision

Assay precision or reproducibility describes the closeness of individual measures of an analyte when the test procedure is applied repeatedly to a sample/control. Types of reproducibility include comparisons among multiple subjects (inter-subject), multiple replicates within one run (intra-assay), run-to-run (inter-assay) and operator-to-operator. Except for replicates within a run, there is no specific CV requirement for the assays discussed in this document; however, the result should be reported in the validation.

To determine the variability of responses among subjects, at least five to ten subjects should be tested. These samples may be obtained from normal donors or from patients or animal models. Reproducibility between plates (inter-plate) (two different plates tested on the same run) can be obtained by testing two to four replicates of each sample/control on two or more different plates on the same run. Reproducibility between runs (inter-day) can be obtained by testing two to four replicates of each sample/control on at least three separate runs on different days. Samples/controls tested on different days should be from identical aliquots. Reproducibility within one run can be obtained by testing four to eight replicates of each sample/control on one single plate. Operator-to-operator reproducibility should be evaluated if more than one operator is expected to perform sample analysis. This can be assessed by comparing the results obtained by at least two different operators testing three replicates of each sample/control over a run.

2.5 Data analysis storage and handling

Data analysis may vary with assay platform and properly validated software must be chosen whenever possible. Before the performance of validation assays, raw data need to be defined based on the particular assay platform and software required. Appropriate storage of the data must be determined taking in consideration relevant Standard Operating Procedures (SOPs) and software capabilities of the instruments used. All data, including raw data, must be archived according to departmental, corporate and government regulations based on where the drug is expected to be registered.

2.6 Acceptance criteria

Upon optimization of the assay, acceptance criteria for the assay are established. While the acceptance criteria may vary with assay platform, some minimal criteria are suggested. Therefore, the results are valid if the following criteria are met:

1. CV must be $\leq 30\%$ for samples and controls. When more than two replicates are done, if the CV for a replicate set is $\geq 30\%$ the two out of three rule, or similar rules, may be used where one of the replicates is an obvious aberrant result and may be eliminated from the calculation. The

definition of an aberrant result will be determined by the assay platform but should be clearly stated.

2. A cut-off value may be used to determine whether positive results are positive and negative results are negative and can be included in the acceptance criteria.
3. When a standard curve is used, the acceptance criterion for the standard curve is that the mean result of each standard sample should be $100 \pm 30\%$ that of the nominal value.
4. If standards and controls have met acceptance criteria, but any one sample has not, only those samples will need to be reassayed.

References

Cristofanilli M, Budd GT, Ellis MJ, Stopeck A, Matera J, Miller MC, Reuben JM, Doyle GV, Allard WJ, Terstappen LW and Hayes DF (2004). *New Engl J Med*, **351,** 781–91.

Cristofanilli M and Mendelsohn J (2006). *Proc Natl Acad Sci USA*, **103,** 17073–17074.

FDA (1994). US Department of Health and Human Services, Food and Drug Administration.

FDA (2001). US Department of Health and Human Services, Food and Drug Administration.

FDA (2006). US Department of Health and Human Serivces, Food and Drug Administration.

Keilholz U, Weber J, Finke JH, Gabrilovich DI, Kast WM, Disis ML, Kirkwood JM, Scheibenbogen C, Schlom J, Maino VC, Lyerly HK, Lee PP, Storkus W, Marincola F, Worobec A and Atkins MB (2002). *J Immunother*, **25,** 97–138.

Kowalski R, Post D, Schneider MC, Britz J, Thomas J, Deierhoi M, Lobashevsky A, Redfield R, Schweitzer E, Heredia A, Reardon E, Davis C, Bentlejewski C, Fung J, Shapiro R and Zeevi A (2003). *Clin Transplant*, **17,** 77–88.

3 Development, optimization and validation of cell-based assays – 2

Manjula Reddy* **and Uma Prabhakar**

Centocor Research and Development, Inc., 145 King of Prussia Road, Radnor, PA 19087, USA

3.1 Introduction

In vitro analysis of cellular immune function provides a qualitative and quantitative measure of cellular responses generated following administration of a therapeutic biologic agent. Cell-based assays cover a variety of functional immune parameters including, but not limited to, cell surface marker expression, cytokine secretion, activation, apoptosis, and proliferation measured by numerous assay/technology platforms. These assays may be used to support safety and efficacy endpoints in a non-clinical or clinical study design. Tests for immunity include *in vitro* functional measures that

* To whom correspondence should be addressed.

Validation of Cell-Based Assays in the GLP Setting: A Practical Guide Editors, Uma Prabhakar and Marian Kelley

require an adequate source of T cells reflective of the immune response to immunization. A variety of immunologic tests are available for monitoring immune responses for vaccine trials in cancer patients (Mosca et al 2005; Keilholz et al., 2002).

One important factor in the conduct of cell-based analysis is to consider the performance characteristic of the assay in detecting immune responses and whether the assay results predict clinical outcome. Desirable assay characteristics include adequate sensitivity, specificity, reliability, reproducibility, simple and rapid to perform, and, in relevant cases, the ability to measure the state of *in vivo* T cell reactivity. There is considerable interest in the use of peripheral blood mononuclear cells (PBMCs) for monitoring the immunological effects in clinical trials. These cells comprise T and B lymphocytes, monocytes, Natural Killer cells and dendritic cells. Collection of PBMCs at the appropriate timepoints permits fresh sample analysis in real time or batch analysis in a central laboratory using cryopreserved samples. Cryopreserved PBMCs have been used for evaluating cellular immune responses in several studies (Weinberg et al, 2000; Christian et al, 2003; Maecker et al, 2005; Kreher et al, 2003; Costanini et al, 2003). A study by Weinberg et al, 2000 indicated that cryopreserved cells are suitable for longitudinal studies of the CMV-specific immune responses in HIV-infected patients and uninfected controls. The use of cryopreserved cells involves collecting the blood samples, and conducting the cryopreservation prior to shipping the cells directly to the testing facility for immediate testing. Alternatively PBMCs can be isolated, cryopreserved and shipped to a central testing laboratory where they are stored for batch testing of samples at a later date. This can be preferable since it facilitates analysis of longitudinally collected patient samples into single assays, thereby limiting variability associated with inter-day assessment.

The ability to analyze cryopreserved PBMCs for antigen specific T cell immunity is required to evaluate responses to immune-based therapies. Likewise, the ability to cryopreserve lymphocytes in PBMCs to retain their function after thawing is critical to the analysis of cancer immunotherapy studies. Several reports emerged in the last few years from investigators who evaluated a variety of cryopreservation strategies with the aim of developing an optimized protocol for freezing and thawing PBMCs to retain viability and function (Disis et al 2006; Betensky et al 2000; Weinberg et al 2000). Cell viability measures were performed to study the effects of altering several key steps in the cryopreservation process such as the volume of the washes, number of frozen cells per tube, media additives, and temperature during the thawing process (Disis et al 2006). They found two key parameters which resulted in an optimal cryopreservation method that preserved both cell viability and antigen specific T cell function as assessed by lymphoproliferation.

Human serum albumin was the optimal media additive for decreasing cryodamage, and diluting thawed cells in a 37 °C media bath maximized cell survival.

This chapter provides specific recommendations for the optimal validation of immunoassays used to evaluate cellular immune responses using PBMCs isolated from clinical or non-clinical peripheral whole blood samples. Additionally, it describes the relevant validation parameters to be addressed based on the intended use of the assay and the procedural documentation required. In addition, any other relevant conditions not listed in the current document should also be assessed based on the application of the assay.

3.2 Definitions

Sample	Any cryopreserved peripheral blood mononuclear cells (PBMCs) or purified lymphocyte population taken from subjects in a clinical or non-clinical study sample
Assay control	Sample from a normal, pre-tested donor treated with the same stimulation agents as the test samples and included on all assay plates and used to monitor the assay outcome (pass/fail)
Positive control	Sample expected to elicit a positive response in the parameter measured as after treatment
Negative control	Blank sample used to monitor background in the assay and usually includes medium without cells and/or cells without stimulating agents
Predose control	Clinical or non-clinical study sample collected prior to any drug treatment
Optimal condition	An experimental condition (concentration of reagents, length of stimulation, etc.) that results in an acceptable signal to noise ratio
SD	Standard deviation is the square root of the variance and is defined by the following formula: $SD = \sqrt{[\sum(x - \text{mean})^2/(N-1)]}$ where x is an individual value and N is the sample size.
CV	Coefficient of variation is an indicator of precision and defined by the following formula: $CV = \text{Standard deviation}/\text{Mean} \times 100$

The procedure for establishing a validated cell-based assay includes assay development, optimization, validation, appropriate data analysis, and storage.

3.3 Assay development and optimization

The immunological monitoring of antigen-specific T cell responses is becoming a commonly used tool for many clinical studies including therapeutic HIV trials (Magee et al, 2004; Dalton et al 2005; Putz et al, 2004; Weinberg et al 2000; Keilholz et al 2002). Testing stored samples accumulated over time from the same donor can potentially minimize within-donor, inter-day (inter-assay), and inter-operator variabilities. Therefore, for PBMC-based assays, it is important to ensure that blood samples are handled in a manner that will not compromise the ability of the cells to respond to activation stimuli, that the viability of the cells is not significantly altered, and that enough cells are recovered to perform the desired assays (Ruitenberg et al, 2006).

Many factors have been shown to affect T cell functional responses including PBMC isolation, shipment, storage, cryopreservation, and thawing (Costantini et al, 2003; Ruitenberg et al, 2006; Betensky et al, 2000; Weinberg et al, 1998; Maecker et al, 2005; Disis et al, 2006; Kreher et al, 2003). Several of these reports suggest methods for improving these influencing factors to preserve and optimize lymphocyte function and viability. Development of cell-based assays using PBMCs should include testing of several assay parameters described in the following sections, that result in optimal responses as defined by an appreciable signal to noise ratio to be able to identify positive response over background response that distinguishes high and low responders.

PBMC isolation from peripheral whole blood samples

There are two methods of PBMC isolation available to clinical investigators: (1) blood collected in Vacutainer tubes and shipped to a central location for PBMC isolation using Ficoll density gradient separation, or (2) blood collected in a Vacutainer CPT™(Cell Preparation Tube) and centrifuged immediately at the collection site and then shipped to a central location for PBMC recovery. A recent study that compared these two methods reported that PBMCs prepared by both methods perform equivalently and show similar yield and viability (Ruitenberg et al, 2006). The use of CPT tubes to prepare PBMC from whole blood samples could be a viable alternative that allows the shipment of samples from multiple sites to a central laboratory. However, assay developers should further examine PBMCs from both these options for assay specific performance. Irrespective of the method of PBMC isolation, the following parameters should be optimized prior to assay validation.

Cryopreservation and thawing of PBMCs

Maintaining cell viability and functionality by using the appropriate cryopreservation and thawing procedures is key to the successful outcome of assays using PBMCs. Freezing with suitable pre-tested freezing media (human AB serum or fetal calf serum with 10% dimethy sulfoxide, DMSO) along with proper thawing procedures (temperature and volume of washing media) should be optimized. No difference in viability have been reported when cryopreserved PBMCs were thawed and resuspended in different volumes (15–50 mL) nor when centrifuged for 5 or 10 minutes, nor when number of cells per vial ranged from 1 to 3×10^7. However, the temperature of media used to wash the cells after removal from the cryotubes was reported to affect viability. Media maintained at 37 °C was found to be optimal (Dissi et al 2006). It has been reported that using room temperature DMSO results in better viability, recovery and functionality compared to the conventional ice cold freezing medium (Kreher et al 2003).

Preserving viability and function without altering the native responses is the goal during the freeze/thaw process. Evaluating preserved stored PBMCs from over 50 donors it was shown that when the viability of thawed cells was >/= 70% a consistent proliferative T cell responses could be expected. The type of protein additive and the media temperature at the time of washing thawed cells appeared to affect the viability and functionality of T cells (Disis et al 2006). In our laboratory, we found that cryopreservation by resuspending PBMCs after isolation at 2×10^6 cells/mL in decomplemented human AB serum freezing media and adding an equal volume of 80% human AB serum with 20% DMSO, pre-warmed to room temperature yielded >90% viability and elicited functional responses in Enzyme-Linked ImmunoSpot (ELISPOT) and activation assays equivalent to non-cryopreserved PBMCs.

Cell density

For assays that are cell density dependent (ELISPOT, proliferation etc), titration experiments should be performed to determine the optimal number of PBMCs required to generate measurable responses over the background. We recommend a range of cell number from 10,000 to 300,000 cells/well.

Stimulation conditions

For all assays, the reagents used for stimulation or activation of PBMCs should be titrated to determine the optimal concentration that provides a good response over background (negative control). Several agents stimulate release of a particular factor (e.g. cytokine) from PBMCs, therefore a stimulating agent that is appropriate and relevant for the factor to be measured should be selected from evaluation of various such agents (Reddy et al, 2004). The stimulating agents should be titrated to determine the optimal concentration that shows the best signal to noise responses that are specific for the analyte being measured.

Cell culture conditions

Optimization of cell culture conditions using appropriate media for the assay and serum should be performed. Appropriate batches of media additives such as serum should be tested and those that produce low background without affecting cell functionality, selected. The length of cell culture should also be determined based on the kinetics of response to the stimulating agents. The time point at which there is acceptable positive response over background should be selected.

Specificity and interference

The specificity assessment should include: (1) specificity of measured response and (2) specificity of reagents used to measure the intended response. The specificity of the response elicited by the stimulation reagents should be assessed using an irrelevant control or a blocking agent and in the presence of other interfering factors such as cytokines and expected concomitant medications. The specificity of assay reagents such as capture and detection antibody pairs should be examined. The specificity of reagents when used as a single reagent as well as in the presence of other reagents (such as, multi-fluorochrome conjugated antibodies in flow cytometric analysis) should be evaluated as well. The unstimulated condition serves as negative specificity control for the stimulating agent.

Effect of different anti-coagulants on the stimulation response

Several anti-coagulants are reported to show different effects on the functional responses induced by various stimulating agents.(Kumar et al, 2000) The anti-coagulants typically in use include potassium EDTA, heparin and acid–citrate–dextrose (ACD). EDTA is the preferred anti-coagulant for applications, where preservation of cellular integrity and monocyte population is important. (Nicholson et al, 1993). These anti-coagulants also showed differential ability to preserve various lymphocyte subsets. Mitogen and antigen specific responses, along with the cell's ability to proliferate in response to various antigens was also compromised (Kumar et al, 2000, Shield et al, 1983, Weinberg et al, 1998, and Nicholson et al, 1993). Therefore, reproducibility of the intended responses should be evaluated in the presence of selected anti-coagulants to choose the one that best preserves the response.

Positive and negative controls

A PBMC-based assay should include at least one positive and one negative control. The response from a negative control (cells without stimulating agent) may be subtracted from all test and positive control results, i.e., used as a blank.

A cell-based assay may have two levels of positive controls and the response expected should be substantially higher than that of the negative control. One type of positive control reagent is usually a polyclonal activator that elicits a response in all samples. Another positive control that serves as an assay control includes a normal pre-tested donor that is batch frozen to be included in all assay plates used for a clinical study. The target response from the positive control donor should be indicative of the assay performance and can also be used to characterize assay sensitivity and specificity.

Stability of responses

The stability of responses should be examined using samples that mimic clinical study sample receipt time (usually following day post-blood draw) and compared to isolated PBMCs tested immediately after blood draw. If the markers or immune responses using PBMCs from overnight-shipped blood samples are not stable PBMCs from overnight-shipped blood samples, they should not be considered for validation. However, if they are stable, assay validation should proceed as outlined below. Additionally, PBMC isolation procedures with Ficoll might affect expression of some markers that otherwise are stable in lymphocytes from whole blood samples, such as CCR5. Also functional parameters such as cytokine or activation responses should be assessed in PBMCs obtained from overnight shipped blood samples and compared to results obtained from the same aliquot of blood sample that was tested immediately after blood collection. Similarly, PBMCs isolated and cryopreserved from blood samples on the same day vs. next day post-collection should be compared side by side in the same assay to assess stability of markers on the immune responses.

3.4 Validation parameters

Once the optimal assay conditions are identified, the validation of a PBMC assay should include the following sections as applicable. A validation plan must be prepared including various validation parameters as described below based on the intended use of the assay.

Stability of the responses

Effect of anti-coagulant

All stability tests should be performed using PBMCs isolated from whole blood samples collected in the presence of the anti-coagulant selected during the assay optimization. As donor-to-donor variability exists, a minimum of five donors should be evaluated for all stability measurements.

Effect of shipment

Samples should be shipped to mimic the anticipated blood sample receipt time and temperature from clinical sites. PBMCs should be isolated from blood samples prior to shipment and after shipment and cryopreserved. These cryopreserved PBMCs should be tested side by side to determine stability of the responses during shipment. Once sample stability is determined, all the subsequent validation parameters should be evaluated using samples that mimic clinical study samples (fresh or shipped). Compare stability of at least ten freshly prepared PBMC samples with those that have been stored under conditions that might be used for clinical trial samples (length of storage and temperature). Common storage conditions include liquid nitrogen storage (for 1–2 years). Normal donor samples or non-study related disease samples with appropriate consent might be used for this analysis.

Sample variability

Inter-subject variability

Inter-subject variability should be determined under similar test conditions using at least ten donors to assess the range of responses in normal donors and if available should also include treatment naive samples representing relevant diseases. Assay precision measures the reproducibility of the assay within several replicates of same sample in a plate, between plates within single assay and between days.

Intra-assay variability

Cryopreserved cells or aliquots of whole blood from the same sample vial should be assayed in parallel in triplicate wells (within-plate) or tubes to determine the intra-assay variability of the assay. A minimum of five normal donor samples should be tested.

Inter-assay variability

Cryopreserved cells from the same sample vial should be assayed in three different assays on the same day. For plate-based assays, three different assays should be set up in three plates using the same test sample to determine the within-day and between plate variability. A minimum of five normal donor samples should be tested.

Inter-day variability is determined by testing three different cryovials (from the same batch) thawed on three separate days and assayed using the same lot of reagents and assay procedures. A minimum of five normal donor samples should be evaluated.

Inter-operator variability

Cryopreserved PBMCs cells from at least five normal donor samples should be assayed in two different experiments by at least two trained operators in triplicate.

Assay robustness

Variability of assay performance with different lots of assay materials such as plates, lots of stimulation agents should be determined. Further, stability of critical assay steps such as the stability of responses prior to instrument read-outs as well as stability of critical assay reagents (e.g. antibody coated plates or pre-diluted reagents) should be determined.

Limits of quantification

Limits of quantification should be established for each assay based on the results obtained from validation experiments. The lower limit of quantification (LLOQ) for measuring cellular immune responses is defined as the lowest measurable response over background for the specific assay format. The upper limit of quantification (ULOQ) is defined by the maximum positive measurable response for the specific assay format.

Acceptance criteria

Using the data obtained from the validation experiments, acceptance criteria for samples, assay, and data must be established for clinical implementation of the assay. The following sections outline some of these criteria that are typically established for PBMC-based functional assays.

Samples

Some cellular pharmacodynamic markers and cellular immune responses are not stable over time after blood draw; therefore, criteria should be set to exclude samples received after a certain time of blood draw. We have observed that PBMC isolation and response is affected by the duration of storage of blood samples over time. Therefore, blood samples received 2 days post-blood draw are excluded from analysis.

Assay

The acceptance of results from a plate-based PBMC assay is defined by the response of the positive control donor sample in the assay plate. If the positive control donor does not meet *a priori* defined responses with the assay positive and negative controls, all samples from that assay plate should be re-tested. Also results from each test sample are accepted based on the responses of positive and negative controls for the sample.

Data

All the samples are tested in triplicates or duplicates as determined by assay precision results. The%CV (range of 20–40%) will be based on the intra-assay validation results and the scientific validity of the responses. Application of the one out of three rule for excluding outliers may be considered. Depending on the rigidity of the assay and based on intra-assay precision results, samples may be analyzed in duplicates (as the cell yield is the usual limiting factor in cell-based assays).

Data analysis, storage and handling

Data analysis varies with assay platform and properly validated software must be chosen whenever possible. Before the performance of validation experiments, naming convention of data files, storage of the raw and analyzed data must be defined taking into consideration relevant standard operating procedures and software capabilities of the instruments used. All data including raw data must be archived according to departmental and corporate regulations.

Assay implementation in clinical studies

In clinical trials, PBMC samples may have to be studied longitudinally with the blood drawn at several time points in order to assess the efficacy of treatment. Ideally, to minimize inter-assay variations, these samples should be tested side by side in a single assay. All related clinical samples must be tested using a single lot of assay reagents and in a batch mode testing to minimize any variability contributed by factors other than the sample itself to be able to accurately determine effects due to treatment. Also each assay plate should include assay controls for monitoring assay performance and a positive and negative control for the samples.

References

Betensky RA, Connick E, Devers J, Landay AL, Nokta M, Plaeger S, Rosenblatt H, Schmitz JL, Valentine F, Wara D, Weinberg A and Lederman HM (2000). Shipment impairs lymphocyte proliferative responses to microbial antigens. *Clin Diagnostic Lab Immunol*, **7**, 759–763.

Christian RK, Markus TD, Robert G, Boehm BO and Tary-Lehmann M (2003). $CD4^{+}$ and $CD8^{+}$ cells in cryopreserved human PBMC maintain full functionality in cytokine ELISPOT assays. *J Immunol Methods*, **278**, 79–93.

Clay TM, Hobeika AC, Mosca PJ, Lyerly HK and Morse MA (2001). Assays for monitoring cellular immune responses to active immunotherapy of cancer. *Clin Cancer Res*, **7**, 1127–1135.

Costantini A, Mancini S, Gioliodoro S, Butini L, Regery CM, Silvestri G and Montroni M. Effects of cryopreservation on lymphocyte immunophenotype and function (2003). *J Immununol Methods*, **278**, 145–155.

Dalton RS, Webber JN, Pead P, Gibbs PJ, Sadek SA and Howell WM (2005). Immunomonitoring of renal transplant recipients in the early posttransplant period by sequential analysis of chemokine and chemokine receptor gene expression in peripheral blood mononuclear cells. *Transplant Proc*, **37**, 747–751.

Disis ML, de la Rosa C, Goodella V, Kuan L-Y, Chang JCC, Kuus-Reichel K, Clayc TM, Lyerlyc HK, Bhatia S , Ghanekar SA, Maino VC and Maecker HT (2006). Maximizing the retention of antigen specific lymphocyte function after cryopreservation. *J Immunol Methods*, **308**, 13–18.

Keilholz U, Weber J, Finke HJ, Gabrilovich ID, Kast WM, Disis LM, Kirkwood MJ, Scheibenbogen C, Schlom J, Maino V, Lyerly HK, Lee PP, Storkus W, Marincola F, Worobec A and Atkins BM (2002). Immunologic monitoring of cancer vaccine therapy: results of a workshop sponsored by the society for biological therapy. *J Immunotherapy*, **25**, 97–138.

Kreher CR, Dittrich MT, Guerkov R, Boehm OB and Tary-Lehmann M (2003). CD4+ and CD8+ cells in cryopreserved human PBMC maintain full functionality in cytokine ELISPOT assays. *J Immunol Methods*, **278**, 79–93.

Kumar P and Satchidanandam V (2000). Ethleneglycol-bis-(β-aminoethylether) tetraacetate as a blood anticoagulant: preservation of antigen-presenting cell function and antigen-specific proliferative response of peripheral blood monuclear cells from stored blood. Clin Diagnostic Lab Immunol. July, **7(4)**, 578–583.

Maecker HT, Moon J, Bhatia S, Ghanekar SA, Maino VC, Payne JK, Kuus-Reichel K, Chang JC, Morse MA, Lyerly HK, DeLaRosa C, Ankerst DP, Disis ML (2005). Impact of cryopreservation on tetramer, cytokine flow cytometry and ELISPOT. *BMC Immunol*, **18**, 6–17.

Magee CC, Womer KL, Khoury SJ and Sayegh MH (2004). Assessment by flow ytometry of intracellular cytokine production in the peripheral blood cells of renal transplant recipients. *Clin Transplant*, **18**, 395–401.

Mosca PJ, Clay TM, Morse MA and Lyerly HK (2005). Immune monitoring. *Cancer Treat Res*, **123**, 369–388.

Nicholson, JKA, Green TA, and collaborating laboratories (1993). Selection of anti-coagulants for lymphocyte immunophenotyping. Effect of specimen age on results. *J Immunol Methods*, **165**, 31–35.

Putz T, Ramoner R, Gander H, Rahm A, Bartsch G, Holt L and Thurner M (2004). Monitoring of CD4+ and CD8+ T-cell responses after dendritic cell-based immunotherapy using CFSE dye dilution analysis. *J Clin Immunol*, **24**, 653–663.

Reddy M, Erikis E, Davis C, Davis HM and Prabhakar U (2004). Comparative analysis of lymphocyte activation marker expression and cytokine secretion in stimulated human peripheral blood mononuclear cell cultures: an in vitro model to monitor cellular immune function. *J Immunol Methods*, **293**, 127–142.

Ruitenberg JJ, Mulder CB, Maino VC, Landay AL and Ghanekar SA (2006). VACUTAINER®CPT™ and Ficoll density gradient separation perform equivalently in maintaining the quality and function of PBMC from HIV seropositive blood samples. *BMC Immunol*, **7**, 11.

Shield, CF, Marlett P, Smith A, Gunter L and Goldstein G (1983). Stability of human lymphocyte differentiation antigens when stored at room temperature. *J Immunol Methods*, **62**, 347–352.

Weinberg A, Betensky R, Zhang Li and Ray G (1998). Effect of shipment, storage, anti-coagulant, and cell separation on lymphocyte proliferation assays for human immunodeficiency virus-infected patients. *Clin Diagnostic Lab Immunol*, **5(6)**, 804–807.

Weinberg A, Wohl DA, Brown DG, Pott GB, Zhang L, Ray MG and van der Horst C (2000). Effect of cryopreservation on measurement of cytomegalovirus-specific cellular immune responses in HIV-infected patients. *J Acquir Immune Defic Syndr 1*, **25**, 109–114.

4 Whole blood *ex vivo* stimulation assay development, optimization and validation

Manjula Reddy* **and Uma Prabhakar**

Centocor Research and Development, Inc., 145 King of Prussia Road, Radnor, PA 19087, USA

4.1 Introduction

From the first report of the *ex vivo* stimulation assay in whole blood (WB) (Kirchner et al, 1982), this technique has proved to be useful for a variety of applications, in the determination of high and low cytokine producers in sepsis patients (Heagy et al, 2000), and for pharmacological assessments of compounds *in vivo* and *in vitro* (Van der Linden et al, 1998). Changes in cytokine profiles have been shown to be predictive of clinical outcome

* To whom correspondence should be addressed.

Validation of Cell-Based Assays in the GLP Setting: A Practical Guide Editors, Uma Prabhakar and Marian Kelley

in selected disease states (Mira et al, 1999; Hyjek et al, 1995; Sousa et al, 1998; Pinto et al, 2006; Noga et al, 2006). Measuring cytokine production can serve as a valuable biomarker tool for the identification of various diseases and for monitoring response to treatment. These assays are proving to be important in supporting efficacy endpoints. However, accurate and reproducible methods for measuring cytokines such as tumor necrosis factor alpha are critical for their application as clinical biomarkers for disease.

WB cytokine release models have been in use for pharmacological *in vitro* and *ex vivo* studies and have been extended to immunotoxicology testing to permit the potency testing of immunostimulants and immunosuppressants (Langezaal et al, 2002). WB cell culture is an ideal *ex vivo* method to analyze cytokine secretion in a controlled environment and shows utility to study the biological effects of immunomodulators such as dexamethasone on Th1 vs Th2 cytokines (Franchimont et al, 1998). Apart from the use of *ex vivo* stimulation assays for measurement of cytokine release, mitogen-induced *ex vivo* WB lymphocyte proliferation is a widely used method to assess lymphocyte responsiveness to immunosuppressive therapy (Magee et al, 2002).

Ex vivo stimulation assays have been used successfully to examine activation and cytokine production of various immune cell populations at the single cell level in WB (Rodriguez-Caballero et al, 2004). Although it is desirable to isolate respective immune cells from blood, it is evident that such isolated cells do not always reflect the *in vivo* physiological environment. Cell isolation procedures often tend to alter/affect cellular responses, they may stimulate the cells, affect the interactions between different cell types, and eliminate plasma components (Schindler et al, 2004). Since WB assays overcome these disadvantages, they are being explored for other functional studies. The WB stimulation assay has been extended for study of peripheral blood analysis of dendritic cell function to evaluate Toll-like receptor function obviating the need for lengthy purification techniques which include *ex vivo* manipulation and opportunities to introduce artifacts (Ida et al, 2006). Apart from applications using fresh or shipped blood, use of cryopreserved WB to test immune function or detect pyrogenic contamination has also been reported (Schindler et al, 2004). Frozen WB was shown to retain sensitivity and functionality regarding stimulation of cytokine release in response to inflammatory agents (Schindler et al, 2004; Pinto et al, 2005). Such cryopreserved WB can be used as an assay standard or positive control to monitor assay performance. Further, the cryopreservation method enables preservation of patient samples for later analysis of cell functions.

When monitoring analytes of interest such as, cytokines with an *ex vivo* stimulation assay, a thorough understanding of the contribution of various assay components to assay variability is essential to be able to distinguish statistically significant differences due to treatment. The majority of factors impacting assay robustness and precision were found to be pre-analytical, including factors such as gender, diet, biorhythms, sample collection procedures and preservatives, sample transport and storage conditions, specimen handling and processing techniques along with assay specificity (Bonini et al, 2002). These variables must be identified and controlled to minimize their impact so that their effects can be distinguished from changes related to clinical efficacy (Ray et al, 2006). It was reported that the total extent of variability contributed by pre-analytical and analytical variables was much less than the variation between donors, indicating that comparison within a donor is more valid than between donors (Van der Linden et al, 1998).

In a clinical drug development setting, majority of the studies are conducted with samples collected at multiple sites, on multiple days, with multiple collections within a day and require overnight shipping of samples. The assay must be optimized and validated such that small but relevant changes due to disease or treatment are detected in these shipped samples. Early evaluation of the WB stimulation system includes the method of sample collection, transport, handling, and the stimulus. Additional evaluation investigates the supernatant storage, processing, and determination of the cytokine production by immunoassay methods such as enzyme linked immunoassay (ELISA) or multiplex. Both stages of the WB stimulation system contribute the total variation in the WB system. The objective of the WB assay development phase should be to establish a validatable assay that shows less variability relative to the intra- and inter-individual variations, to be able to detect clinically relevant changes in immune responses. The following sections in this chapter discuss the relevant assay variables that should be considered for optimization and validation to evaluate cellular immune responses using WB samples obtained from clinical study subjects. This chapter provides a general overview of these validation parameters, but the actual level of validation performed may be modified based on the intended use of the assay.

4.2 Definitions

Quality control sample	A control that reflects the study sample as closely as possible and should be included in each assay (if possible) to assess the validity of the assay run.
Sample	WB pre-clinical or clinical study sample
Positive control	An assay sample provided by the stimulation of interest (such as stimulation with a polyclonal activator) that elicits maximum response from an *in vivo* drug treated or *ex vivo* stimulated sample.

Negative control	A sample that reflects the background in the assay. It is usually an *in vivo* drug treated sample treated *ex vivo* with all reagents used for the positive control sample except the stimulating agent/s.
Predose Control	Clinical or non-clinical study sample that is collected prior to any drug treatment and *ex vivo* stimulated.
Assay control	A normal pre-tested donor to be included on all assay days (if feasible) and should indicate the assay outcome (pass/fail). This positive control can also be used to characterize assay sensitivity and specificity and can be a cryopreserved WB sample.
Optimal condition	An experimental condition (concentration of reagents, length of stimulation etc) that results in an acceptable signal to noise ratio and indicates a positive response over assay background.

4.3 Assay development and optimization or qualification

Development of *ex vivo* stimulation assays should include testing the following assay parameters to determine the optimal responses as applicable for the intended use of the assay.

Blood collection and effect of anti-coagulants

The importance of proper blood collection procedures and use of different anti-coagulants were discussed in a report by Vaught (2006). Several types of anti-coagulants that differ in their mechanism of action are reported to show different effects on the functional responses induced by various stimulating agents (Vaught, 2006). Diurnal variation and day-to-day variability can be due to the type of anti-coagulant used rather than due to the response of the sample itself. Therefore, stability of the responses of interest should be evaluated in the presence of various anti-coagulants and one that best preserves the response is selected. The *ex vivo* production of the analyte of interest should be determined with at least five subjects using blood samples collected in Vacutainer tubes that are sterile and endotoxin-free, containing different anti-coagulants (heparin, potassium EDTA, or sodium citrate). Additionally, the evaluation of type of sample collection and inclusion of a placebo group is important.

It was reported by Ray et al (2006) that sodium citrate anti-coagulant showed a more consistent kinetic profile at cold temperature and was suitable for shipping and storage of samples collected at multiple clinical locations. While citrate based anti-coagulant was found to be less variable, each application of the assay should select appropriate anti-coagulant for identifying the specific analyte of interest.

Stimulation conditions (stimulation agents, concentration, temperature, length of stimulation and stimulation tube)

Agents that stimulate production of the analyte of interest should be evaluated at different concentrations. Lipopolysaccharide and phytohemagglutinin A are frequently used as powerful stimulants of cytokine secretion, mainly by monocytes and lymphocytes respectively. The appropriate length of stimulation that is feasible (taking into consideration the sample viability) and one that excludes non-specific effects should be determined. For longer stimulation conditions, dilution with media that preserve the functionality of WB cells should be tested.

The stimulating agents should be titrated to determine the optimal concentration that shows the best signal to noise response in the shortest stimulation time.

Responses from *ex vivo* stimulation assay should be evaluated in blood samples aliquoted into replicates of both silanized and non-silanized borosilicate tubes. It was reported that silanized tubes improved reproducibility by reducing the binding of monocytes and/or cytokines to the wall of the tube (Ray et al, 2006). The appropriate temperature of stimulation (usually 37 °C) and suitable equipment that best maintains this temperature, such as, 37 °C water bath, dry heating device, dry heat 37 °C CO_2 incubator should also be evaluated.

Specificity and Interference

The specificity of the response elicited by the stimulation reagent should be ascertained using an irrelevant control or a blocking agent. Further, the detectability of the analyte resulting from the stimulation response should be assessed in the presence of other interfering factors such as other cytokines and therapeutics that are likely to be present in the test samples.

Positive and negative controls

Identification of the appropriate controls of WB assay is vital to the interpretation of the assay results. WB assay controls should include at least one positive and one negative control. The response from a negative control (samples without stimulating agent) may be subtracted from all test and control results, i.e., used as a blank. A positive control reagent is usually a polyclonal activator that elicits a response in all drug treated samples. The response of the positive control reagent should be substantially higher than that of the negative control. A control sample (e.g. a normal pre-tested cryopreserved WB sample) can be included on all assay days to monitor the consistency of assay performance.

Effect of sample storage temperature on stability of responses

Effect of storage temperature on stability of the cytokine production should be evaluated.

The time and temperature of sample shipment may affect the ability of the cells to produce the cytokine of interest and both parameters should be investigated. Aliquots of WB samples from at least three normal human donors should be tested immediately and after storage at ambient and at 4°C overnight and up to 48 hours (depending on anticipated time frame of sample receipt). Effect of cold temperature storage on the analyte being measured should be determined.

Diurnal variation in responses

Decrease in the cells' responses to stimulation may occur as the time between sample draw and sample processing increases. Therefore, stability of the responses in samples collected at different times of the day should be assessed for diurnal variability (time intervals selected should mimic anticipated time frame of clinical sample testing).

WB stored for various lengths of time (0 to 6 or 12 hours) after blood collection from at least five healthy volunteers should be examined for variation in responses. All supernatants should be analyzed in a single batch in an appropriate immunoassay platform (ELISA or multiplex analysis). Results are compared to samples at 0 hour (immediately post-blood draw) and the stableness of the response is determined.

Effect of shipment on stability of responses

The stability of responses should be examined in samples that mimic clinical study sample receipt (usually the following day post-blood draw) and compared to samples tested immediately after blood draw. If the analysis will be performed the same day of the blood draw, stability of the responses should be examined at different intervals of the day to assess diurnal variation. Samples should be shipped at the appropriate temperature determined in sample storage experiments (section 1.5) at ambient or at cold temperatures, though it is believed that shipping with ice packs can be more tightly temperature controlled than ambient samples If shipping without cold packs, ambient shipping container should maintain a constant temperature which would be difficult for 24 hours in winter time. Temperature fluctuations during transit can be controlled by the proper insulation/refrigeration materials described in a report by Vaught (2006). For each assay it must be determined if the blood is stable under the planned storage conditions (length of time and temperature).

4.4 Validation parameters

Once the assay format is optimized as described in the section 1.0 the assay method should be defined in a method or standard operating procedure. This defined assay is now ready for validation which assesses primarily the assay robustness and precision. A validation plan should be prepared including various validation parameters as described below based on the intended use of the assay.

Stability of the responses

Effect of anti-coagulant

To evaluate the effect of anti-coagulant on the stability of the cellular or analyte response the WB assay should be performed using WB samples collected from normal human donors in the presence of the anti-coagulant selected during the assay optimization. Including WB samples from any disease subjects is also recommended. As donor-to-donor variability exists, a minimum of five donors of each type (normal and disease) should be evaluated.

Effect of shipment

The shipment conditions to be validated should mimic the anticipated sample receipt time and temperature from clinical sites. All samples should be tested prior to shipment and results compared to those after shipment or storage to determine stability of the responses (or analytes) during shipment.

Assay robustness

Assessment of critical assay steps evaluates the robustness of the assay to expected or unexpected changes.

The stability of responses prior to instrument read-outs as well as stability of critical assay reagents (e.g. stimulation reagents, positive, negative and specificity reagents etc.) should be determined. The consistency of the response to the pre-determined critical reagents, stimulation time etc. over several assays should be evaluated and the %CV reported to demonstrate the method's reproducibility in the light of small perturbations.

Sample variability

The *ex vivo* cytokine production within a single donor varies over time as a result of biological mechanisms and disease. The intra-donor variation can be determined when the WB stimulation is repeated over time with blood samples from one donor. The variation between donors or inter-donor variation can be determined when samples from several individuals are tested simultaneously under similar conditions.

Intra-subject variability

As cytokine responses may vary over a period of time in the same subject, intra-subject variability should be evaluated using at least ten normal donor samples. Diurnal variation and long term variation over few weeks (to mimic anticipated time points in a clinical study) should be assessed. The subjects selected should represent general population, should be a mix of ages and genders. Determining intra-subject variability in treatment naïve disease subjects as well as normal donors is highly recommended. Ideally ten normal donors (five male; five female) [and perhaps ten naïve disease donors] with samples collected over several hours (diurnal) and several days (time points of study) should be tested. Report the intra-subject %CV over time for each parameter assessed for each donor.

Inter-subject variability

Inter-subject variability should be determined under similar test conditions using at least ten donors to assess the range of responses in normal donors and untreated patient samples if available. Since the levels of different analytes vary between males and females, particularly the cytokines, analysis of gender variability should also be included. Using the same ten donors above (five male; five female) assess the %CV overall for all donors.

Assay precision

Precision in WB assay should be determined within replicates in an assay, between assays, and between operators.

Intra-assay variability

Aliquots of WB from the same sample tube should be assayed in parallel in several replicates (three to six) to determine the intra-assay variability of the assay. At least five normal donor samples should be tested. Determine the %CV within a sample for each donor.

Inter-assay variability

Aliquots of WB from the same sample vial should be assayed in three different assays on the same day to determine within-day inter-assay variability. Each assay requires preparation of stimulation and other assay reagents independently. For plate-based assays, three different assays should be set up in three plates using the same test sample to determine the within-plate variability. At least five normal donor samples should be tested.

Inter-day variability was assessed by performing WB stimulation in five healthy subjects on three consecutive days. If the response of WB assay is cytokine release, the supernatants are analyzed in a single ELISA batch in duplicates. The mean *ex vivo* cytokine responses on the first day are compared to those for the other two days. Report of the %CV for each inter-assay parameter tested.

Inter-operator variability

Aliquots of WB from at least two normal donors should be assayed in triplicates by at least two trained operators in two different experiments to determine inter-operator variability. Determine the inter-operator %CV.

Lot-to-lot variability of reagents

Variability of assay performance with different lots of assay materials such as stimulation agents should be determined. This assessment continues throughout the life cycle of the assay and variability is monitored to assure that there is no detrimental affect introduced by using different lots of critical reagents.

Limits of quantification

Limits of quantification of the assay should be established using the results from the relevant validation experiments. The lower limit of quantification for measuring cellular immune responses is defined as the measurable response over background. The upper limit of quantification is defined as the maximum positive accurate measurable response for that specific assay format.

Acceptance criteria

Samples

Some cellular pharmacodynamic markers are not stable over time after blood draw, therefore, criteria should be set to exclude samples received a certain time after blood draw.

Assay

The acceptance criteria for the assay are defined by the response of the positive control donor samples. Positive response of the normal control donor indicates successful assay performance and therefore all the subjects tested in that assay run are accepted. If the positive assay control fails to respond, the test samples would require re-testing. For assays where use of positive donor sample may not be feasible, acceptability of results from test samples should depend on the response from the positive control reagent.

Data/Documentation

Due to the inherent variability of WB assays, samples are tested in more than one replicate, usually in triplicates. The %CV acceptability should be based on the validation and the scientific validity of the responses. Application of the one out of three rule for excluding outliers can also be considered. Depending on the rigidity of the assay and based on intra-assay results, samples can be analyzed in duplicates.

Finally the validation experiments should be compiled into a formal report to establish the validity of the assay method.

Clinical testing of samples

Minimizing assay variation is critical to be able to distinguish the high and low responders for a particular analyte of interest. All clinical samples should be shipped to a central testing laboratory using identical shipping conditions. Assay variation can be minimized by performing all analyses of an experiment with the same batch of reagents. Also, it is critical for the duration of stimulation to be fixed to maximize the uniformity of the procedure. Determination of analyte levels in the stimulated WB supernatants, should be performed in a batch mode to further minimize variability in the later part of the WB assay.

4.5 Summary

Ex vivo stimulation assays are of interest for drug development and are being used for various purposes including assessing the efficacy of compounds *in vivo*. *Ex vivo*, cell-based assays allow assessment of various factors related to drug activity that are not accessible to most *in vitro* assays that include cellular functions, mechanism of action etc. Tight control of cell stimulation conditions reduces variability. Most important of these are incubation temperature, surface properties of the stimulation container, sample storage temperature and the type of anti-coagulant used. By following these suggestions a WB assay conducted in a good laboratory practice setting can prove to be a valuable tool for assessing effects of therapeutic agents on cellular immune responses.

References

Bonini P, Plebani M, Ceriotti F, Rubboli F (2002). Errors in laboratory medicine. *Clin Chem.* 48(5), 691–8.

Franchimont D, Louis E, Dewe W, Martens H, Vrindts-Gevaert Y, De Groote D, Belaiche J and Geenen V (1998). Effects of dexamethasone on the profile of cytokine secretion in human whole blood cell cultures. *Regul Pept*, **73**, 59–65.

Heagy W, Hansen C, Nieman K, Cohen M, Richardson C, Rodriguez JL and West MA (2000). Impaired *ex vivo* lipopolysaccharide-stimulated whole blood tumor necrosis factor production may identify "septic" intensive care unit patients. *Shock*, **14**, 271–276.

Hyjek E, Lischner HW, Hyslop T, Bartkowiak J, Kubin M, Trinchieri G, Kozbor D. (1995). Cytokine patterns during progression to AIDS in children with perinatal HIV infection. J Immunol, 155(8), 4060–71.

Ida AJ, Shrestha N, Desai S, Pahwa S, Hanekom AW and Haslett A.J.P (2006). A whole blood assay to assess peripheral blood dendritic cell function in response to Toll-like receptor stimulation. *J Immunol Methods*, **310**(1–2), 86–99.

Kirchner H, Kleinicke C and Digel W (1982). A whole-blood technique for testing production of human interferon by leukocytes. *J. Immunol. Methods*, **48**, 213–219.

Langezaal I, Coecke S and Hartung T (2001). Whole blood cytokine response as a measure of immunotoxicity. *Toxicol In Vitro*, **15**, 313–318.

Langezaal I, Hoffmann S, Hartung T and Coecke S (2002). Evaluation and prevalidation of an immunotoxicity test based on human whole-blood cytokine release. *Altern Lab Anim*, **30**, 581–595.

Magee MH, Blum RA, Lates CD and Jusko WJ (2002). Pharmacokinetic/pharmacodynamic model for prednisolone inhibition of whole blood lymphocyte proliferation. *Br J Clin Pharmacol*, **53**, 474–484.

Mira JP, Cariou A, Grall F, Delclaux C, Losser MR, Heshmati F, Cheval C, Monchi M, Teboul JL, Riché F, Leleu G, Arbibe L, Mignon A, Delpech M, Dhainaut JF (1999). Association of TNF2, a TNF-alpha promoter polymorphism, with septic shock susceptibility and mortality: a multicenter study. JAMA, 282(6), 561–8.

Noga O, Hanf G, Brachmann I, Klucken AC, Kleine-Tebbe J, Rosseau S, Kunkel G, Suttorp N, Seybold J (2006). Effect of omalizumab treatment on peripheral eosinophil and T-lymphocyte function in patients with allergic asthma. J Allergy Clin Immunol, 117(6), 1493–9.

Pinto LA, Trivett MT, Wallace D, Higgins J, Baseler M, Terabe M, Belyakov IM, Berzofsky JA and Hildesheim A (2005). Fixation and cryopreservation of whole blood and isolated mononuclear cells: Influence of different procedures on lymphocyte subset analysis by flow cytometry. *Cytometry Part B (Clinical Cytometry)*, **63B**, 47–55.

Pinto RA, Arredondo SM, Bono MR, Gaggero AA, Díaz PV (2006). T helper 1/T helper 2 cytokine imbalance in respiratory syncytial virus infection is associated with increased endogenous plasma cortisol. Pediatrics, 117(5), 878–86.

Ray CA, Dumaual C, Willy M, Fill J, O'Brien PJ, Gourley I, Devanarayan V and Konrad RJ (2006). Optimization of analytical and pre-analytical variables associated with an *ex vivo* cytokine secretion assay. *J Pharm Biomed Anal*, **41**(1), 189–195.

Rodríguez-Caballero A, García-Montero AC, Bueno C, Almeida J, Varro R, Chen R, Pandiella A, Orfao A (2004). A new simple whole blood flow cytometry-based method for simultaneous identification of activated cells and quantitative evaluation of cytokines released during activation. Lab Invest, 84(10), 1387–98.

Schindler S, Asmus S, Aulock VS, Wendel A, Hartung T and Fennrich S (2004). Cryopreservation of human whole blood for pyrogenicity testing. *J Immunol Methods*, **294**, 89–100.

Sousa AO, Lee FK, Freiji R, Lagrange PH, Nahmias A (1998). Human immunodeficiency virus infection alters antigen-induced cytokine responses in patients with active mycobacterial diseases. J Infect Dis, 177(6), 1554–62.

Van der Linden MW, Huizinga TW, Stoeken DJ, Sturk A and Westendorp RG (1998). Determination of tumour necrosis factor-alpha and interleukin-10 production in a whole blood stimulation system: assessment of laboratory error and individual variation. *J Immunol Methods*, **218**, 63–71.

Vaught BJ (2006). Blood collection, shipment, processing, and storage. *Cancer Epidemiol Biomarkers Prev*, **15**, 1582–1584.

5 Immunohistochemistry assays in Good Laboratory Practice studies

Frank Lynch[1], Steve Bernstein and Hector Battifora[2]

[1]*QualTek Molecular Laboratories, Newtown, PA, USA*
[2]*QualTek Molecular Laboratories, Santa Barbara, CA, USA*

5.1 Introduction

Brief overview of an immunohistochemistry assay

Immunohistochemistry (IHC) assays may be used in clinical studies when tissue samples are available and scientists and clinicians want to evaluate the expression of a particular target within the sample. Unlike 'grind and find' assays, the IHC assay allows for microscopic review of the tissue histology and evaluation of the specific target's staining pattern on the same tissue section. Thus, the expression pattern of a biomarker may be assessed with cell specific and often intracellular compartment resolution. The IHC staining pattern observed is based on binding of a specific primary antibody. This antibody is detected with an enzymatic colorimetric detection system that enables the visualization of the bound antibody in the

Validation of Cell-Based Assays in the GLP Setting: A Practical Guide Editors, Uma Prabhakar and Marian Kelley

context of the overall tissue and cellular architecture. This is particularly important because tissues are generally heterogeneous and composed of many different cell types. For example, in oncology studies samples may contain tumor, the normal tissue counterpart, infiltrating lymphocytes and the supporting stroma. Core needle tumor biopsies are often obtained in oncology studies and it is not uncommon that little tumor and sometimes no tumor be present in the biopsy.

IHC allows the analysis of which cells, based on their morphology, label for the target protein and also whether tumor, based on morphology and immunoreactivity, is present in the sample. Subcellular localization may also be readily discriminated. A particular target or biomarker may be localized to different compartments of the cell; such as the nucleus, the cytoplasm,

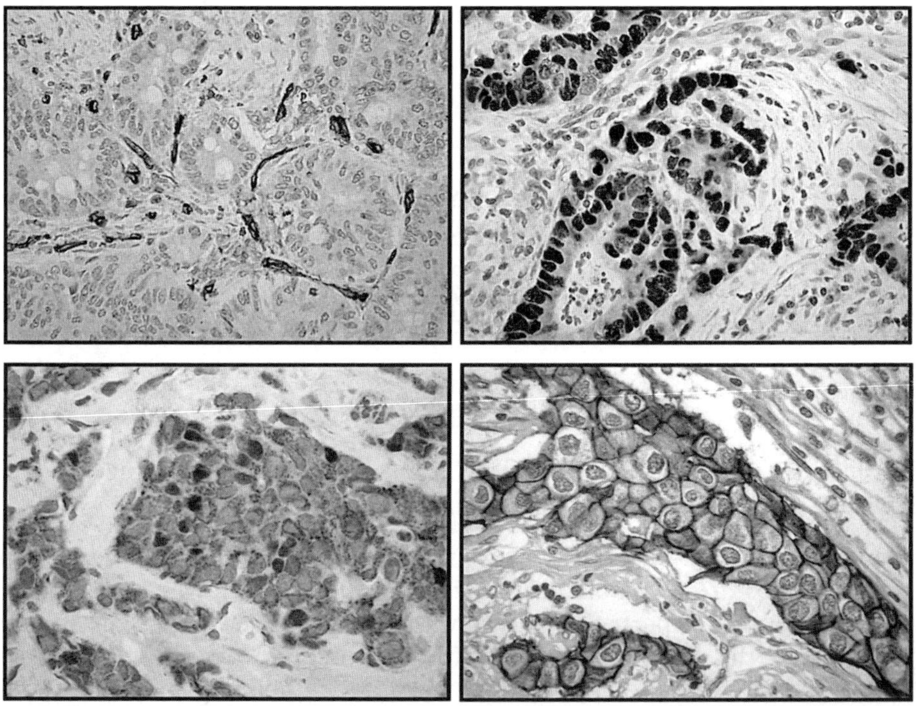

Figure 5.1 Various examples of IHC staining. **Top Left:** CD31 IHC in a colon carcinoma. The CD31 antibody labels endothelial cells of blood vessels. Tumor cells do not label. **Top Right:** p53 IHC in a colon carcinoma showing strong nuclear localization in tumor cells. **Bottom Left:** dUTPase IHC in a breast carcinoma showing both cytoplasmic (mitochondrial) and nuclear localization in tumor cells. **Bottom Right:** HER2 IHC in a breast carcinoma showing strong plasma membrane localization of tumor cells. Positive staining is indicated by the presence of the dark brown chromogen. Hematoxylin counterstain. A full-colour version of this image appears in the colour plate section of this book.

plasma membrane or extracellular (see Figure 5.1). Localization within the nucleus and cytoplasm may further be delineated depending upon their pattern of staining, i.e., diffuse staining, fine or coarse granular staining, focal staining, perinuclear staining, etc. The relative intensity of staining of various cell types may also be discerned and may correlate well with the amount of target molecule present, particularly when sample acquisition is highly controlled (Riera et al, 1999). Assays that measure RNA cannot provide this level of information and are particularly vulnerable to sampling errors.

In most cases the primary antibody is not labeled directly, rather a secondary labeled antibody is used to bind the primary antibody. The label on the secondary antibody is often biotin, but other non-biotin conjugates such as fluorescein isothiocyanate (FITC) or newer polymer systems may also be used. The detection system is layered, which allows for an amplification of the signal for better sensitivity. Briefly, tissue sections are prepared from either frozen or paraffin-embedded tissues. Formalin-fixed, paraffin-embedded (FFPE) tissues will be described here since they are more commonly used in standard pathology analyses and offer superior morphology when compared to frozen sections. A tissue sample is typically fixed in 10% neutral buffered formalin (NBF) and then processed and embedded into paraffin so that it can be sectioned. Sections 4–5 μm thick are prepared using a microtome. Fortunately, due to their thickness, many sections, and thus many IHC assays, can be tested even from small specimens. The sections are then placed on glass microscope slides for IHC staining. The sections are deparaffinized in xylene and hydrated through graded ethanols to distilled water. If tissue pretreatment (to be discussed later) to promote antibody binding is required it is performed at this point. After tissue pretreatment, the slides are placed in phosphate-buffered saline (PBS) and then blocked for common immunoglobulins in epitope-free immunoglobulin containing serum followed by incubation with the primary antibody. Primary incubation times may vary from 30 to 60 minutes to overnight incubation to enhance binding. PBS rinses are performed followed by incubation of a labeled secondary antibody, which binds to the primary antibody. For example, if the primary antibody is a mouse monoclonal antibody, the secondary is an anti-mouse antibody conjugated to biotin or some other label or enzyme. PBS rinses are performed followed by incubation with HRP-conjugated avidin. The horse radish peroxidase (HRP) colorimetrically reacts with a chromogen such as 3′3′- diaminobenzidine (DAB), which in the presence of hydrogen peroxide yields a dark brown, permanent, insoluble precipitate. (There are multiple enzyme detection systems and chromogens. The investigator should decide the optimal detection system and chromogen for the proposed study.) Slides are then counterstained in hematoxylin or another dye to allow for the evaluation of tissue morphology and then permanently coverslipped after dehydration through alcohols and xylenes.

In addition to providing cell specific staining, valuable post-translational modifications may also be detected with some IHC assays. An increasing number of phospho-specific antibodies are now commercially available to detect specific phosphorylated amino acid residues on target proteins. Detecting the phosphorylated form of a target protein may be much more valuable than simply detecting the target itself without information regarding its phosphorylation status, since it often indicates activation of the target. This is relevant in clinical studies where therapeutic intervention affects phosphorylation status in specific targeted pathways. Thus, pre- and post-treatment samples may be IHC tested to determine whether the phosphorylation status changes, which may determine that the therapy is working from a mechanistic perspective (Albanell, 2002). Changes in subcellular localization may accompany phosphorylation status in some proteins and provide further evidence as to whether a protein is active or not. For example, phosphorylated targets such as the signal transducer and activator of transcription (STAT) family of proteins label with a nuclear localization; whereas the non-phosphorylated proteins localize in the cytoplasm (Darnell, 1997).

IHC assays have been used extensively in hospital clinical laboratories to assist in the diagnosis and prognosis of patient tissue biopsies. These assays are particularly common in oncology patients since pathological diagnosis often requires surgical removal of tissue samples. IHC assays are particularly important in breast cancer studies. The American Society of Clinical Oncology (ASCO) recommends IHC testing for estrogen receptor (ER), progesterone receptor (PR) and HER2 for all breast carcinomas (ASCO, 1996). These assays help determine the prognosis of the patient and also assist in determining which therapeutic intervention will be used, particularly in targeted therapies such as Herceptin™. HER2 positive patients may be eligible for Herceptin™ therapy based on the IHC staining pattern. The HER2 assay is a prime demonstration of the importance of IHC assays in the hospital setting (see Figure 5.1). IHC assays have become more standardized with the advent of automation, which provides increased reproducibility and throughput. In addition, more and better characterized antibodies are available to help demonstrate protein expression and cellular changes with new targeted therapeutics.

The laboratory chosen to perform IHC testing for clinical studies is critical and should be chosen according to its capabilities. The laboratory is required to have appropriate standard operating procedures (SOPs) in place and to have the ability to follow these SOPs. Appropriate, well-trained personnel are required along with experienced study directors. The laboratory must understand the project requirements and be able to offer suggestions to enhance the study based on knowledge of how best to use

IHC within a specific study. Knowledge of the history of the laboratory is essential in understanding whether the organization will yield quality results.

The capacity for IHC assays to detect protein, post-translational modifications, subcellular localization and specific cell types underlines the value of these assays in clinical studies. IHC assays are important in clinical studies where investigators want to measure the expression of a biomarker(s) in tissue samples to help understand whether a therapy is working or to help better define the patients potential to respond to therapy. Tissue samples may include diagnostic blocks, pre- and post-treatment specimens and surrogate tissues. This chapter focuses on the general requirements for these assays, particularly in the handling of clinical specimens, IHC assay development and IHC testing and analysis of samples from clinical study subjects.

Herceptin™ and HercepTest™ are trademarks of Genentech.

5.2 Samples for IHC analysis in good laboratory practice studies

Sample handling is critical in good laboratory practice (GLP) clinical studies. Here we focus on the types of samples that are used in these studies as well as the handling of these samples. Quality control and proper tracking and care for these samples are of paramount importance. These samples are truly one of a kind and irreplaceable. Companies invest a great deal of resources into obtaining these samples for the purpose of understanding disease state for the potential purpose of determining therapy or understanding if the therapy is working. More importantly, the individuals being treated in the study also invest in agreeing to contribute these tissues, which in some cases is not a trivial procedure. In many oncology and immune related clinical studies baseline tissue samples are obtained along with one or multiple post-treatment samples. These samples are subsequently tested with various biomarker assays. Before this testing can occur the tissues must first be collected, labeled, shipped, processed (if necessary) and stored all the while maintaining control over the source, type and integrity of the specimen. All these procedures, if improperly carried out, may affect the subsequent morphology and immunostaining of the samples. This subchapter focuses on the requirements of handling these specimens.

When planning a clinical study that requires tissue samples for IHC testing and analysis, several factors need to be considered to maximize the potential of these samples before the study starts. One must first determine which tissue-based assays are required from these samples and then ascertain the best approach for acquiring the specimens. In the preferred scenario the assays to be used should be already in place and the reagents and control materials required for these assays identified and sourced. The number and types of assays will dictate the specific fixation regimen and size requirement of each sample.

The types of assays may dictate the type of fixative required by the tissue. In most instances, tissues will either be fixed in 10% NBF or frozen. Formalin-fixed tissues have many benefits over frozen tissues. Formalin-fixed samples are easier to handle and when embedded into paraffin yield excellent morphology. In addition, FFPE tissue is easy to store, the assays are usually more robust than frozen sections, and the sample maintains its integrity without the need for special storage, such as deep freezers. Most studies utilizing IHC assays will use FFPE tissue samples, which are described here. Specimen containers pre-filled with 10% NBF are commercially available in various sizes to accommodate most biopsies. A general rule is that the volume of 10% NBF should be at least 10 times the volume of the tissue sample. Very small vials should be avoided as they are difficult to handle and can cause problems removing the samples. Tissues should be limited in thickness to about 5 mm to allow for complete fixation of the tissue. Samples thicker than 5 mm should be carefully dissected with a sharp scalpel to meet this requirement.

Over-fixation may be problematic for some IHC assays. Whenever possible, fixation in 10% NBF should be standardized for a study. In a preferred scenario, tissues should be fixed for a set-time period – 24 hours at room temperature is a good standard. In many cases, however, tissue samples may stay in 10% NBF for extended periods of time before being embedded into paraffin. This may be due to a number of factors, including: the samples are processed in separate labs from the sites where they are collected, extended shipping times, and willingness for the site to handle the specimens with special instructions. Normally it is preferred by most sponsors to ask the sites to handle the samples as little as possible to limit potential problems in sample integrity or loss. Fortunately, due to better IHC pretreatment (so called Antigen Retrieval) regimens, over-fixation is now less of an issue for a large proportion of antigens. In addition, effects of extended fixation may also be determined in advance during assay development to ensure that the antigens to be detected are unaffected by prolonged exposure to the fixative (Battifora, 1991).

Formalin-fixed tissue samples are used in most immuno-localization studies; however, other types of samples and fixation procedures may also be used. These sample types include frozen sections and cytology specimens. Frozen sections should be snap frozen and stored at –70 °C. Cytology specimens, either cell blocks, smears or cytospins, may require specific handling and fixation procedures. In addition, previously cut slides of tissue samples may be provided in lieu of a surgical tissue sample although this is not the preferred method.

Labeling and shipping of samples

Each sample should have pre-printed labels with unique identifiers to appropriately match the sample with its donor. These labels should be placed on the sides of the specimen container. Subject information required on the label should be determined in advance. This information usually includes the study protocol number, unique identifier/accession number, the subject's initials and date and time the sample was taken. Labeling information should be on the specimen cup written legibly with an indelible marker that will not run, smear or wipe off, particularly when in contact with the shipping fluid. Labeling information should never be placed on only the specimen cap since caps can be separated from the specimen and are then easy to mix-up with other samples. For formalin fixation, the sample simply needs to be submerged in the 10% NBF solution as soon as possible after removal from the subject. Directions for sample handling should be clearly stated in manuals provided to participating sites by the sponsor or Contract Research Organization.

Specific shipping instructions should also be provided to participating sites. Samples may be shipped from the site to the sponsor, a central laboratory or a specialty laboratory identified by the sponsor. Manifests should accompany all samples and when possible the site receiving the sample should be notified in advance. Care must be taken to make certain the manifests match the sample labels being shipped. They should be clearly marked with the sponsor and protocol number so that the receiving site can readily identify the sample. Samples should be packed securely and temperature-sensitive material should be packaged accordingly. In most cases, tissues being shipped in 10% formalin may be shipped at room temperature. If paraffin tissue blocks are being sent in the summer months a cold pack should be added to the shipping container.

Receiving clinical study samples

The laboratory receiving the clinical study samples should have SOPs in place regarding the receipt of materials for studies conducted in accordance with GLP. Briefly, the receiving laboratory should examine the shipping label and then unpack the shipping container in a defined quarantined receiving area. The contents of the package should be examined and matched to any accompanying manifest or packing list. Each sample should be entered into a receiving log and notes taken regarding the type, size and condition of the sample. For example, if a skin punch biopsy sample is received a notation may be made indicating the size of the punch biopsy in millimeters. Each sample label and written identifications should also be matched to the manifest and any discrepancies should immediately be sent to the sponsor, the project study director and any other previously determined contact personnel. Each sample should then receive a laboratory accession number that clearly identifies it with the clinical protocol, sponsor, subject identifier, treatment and in-house study number.

If samples are received in 10% formalin and require paraffin embedding, the samples should be submitted to appropriate personnel along with the appropriate paperwork. Samples are normally placed on a tissue processor and then embedded into molds that are clearly labeled with an accession number and in-house project or sponsor protocol number. Depending on the types of sample, specific considerations may be required for proper orientation of the samples. For example, skin biopsy samples should be embedded such that the sample is in cross-section upon cutting. After embedding, the samples should be safely stored in designated areas. In some cases, previously embedded tissue blocks or pre-cut unstained sections on slides are provided. These samples should be received according to the laboratory SOPs and safely stored.

Sample storage and chain of custody

Samples should be stored in defined areas consistent with their requirements. FFPE tissues should be placed into designated areas within a heat resistant safe in order to protect the specimens from damage. These samples should only be removed by authorized study personnel and should only be treated according to the previously defined assay requirements. Initial sections of specimens cut onto glass slides and stained with hematoxylin and eosin (H&E) may be required to assess quality and orientation of the sample. Further tissue sectioning should be performed just prior to IHC staining of the samples. Tissues sectioned onto glass slides may lose signal with certain target proteins if stored too long. This is true especially with target proteins that localize to the cell nucleus. Tissue samples should not be shipped from the IHC laboratory unless specifically instructed by authorized sponsor personnel. The IHC laboratory should maintain constant documentation of the location and movement of the clinical study samples. Any long-term storage requirements (longer than 12 months after the last sample is received) should be discussed with the sponsor. It is also important to note that enrollment periods for some clinical studies may span years and not months. As a result, strict policies in the form of SOPs should be in place to keep track and care of the samples.

5.3 IHC assay development

Assay development is perhaps the most challenging aspect of performing IHC assays in studies conducted in accordance with GLP. Assay development requires expertise and can be a time consuming process requiring a great number of slides and assay conditions to be tested. The validated assay relies on availability of specific antibodies that are reactive in tissue sections. The resulting assay should reproducibly label in control and target tissues. Assay development considerations in FFPE tissues include: choice of antibodies, tissue pretreatments, target tissue types and detection systems. The ultimate application of the assay needs to be considered at the onset of testing as this may dictate the approach and extent of testing. For example, an assay that may be eventually used as a companion diagnostic may require further validation, documentation and reporting than early or smaller clinical studies.

Choice of antibodies

For newly developed assays, more than one antibody, if available, should initially be tested per target. Testing multiple antibodies not only increases the chances of identifying the best FFPE reactive antibody but also demonstrates that the target is being correctly identified when two different antibodies provide similar staining patterns. However, antibodies generated to the same target may yield very different staining patterns. This is true of antibodies that are commercially available or novel. Antibodies can be monoclonal or polyclonal. When using polyclonal antibodies, ensure that they are immuno-affinity purified against the target antigen. Simple immunoglobulin purification (e.g., Protein A) is usually not sufficient and can contribute to background due to non-specific binding. Some commercial sources may indicate whether an antibody has been tested for IHC reactivity; however, the majority of antibodies are not sufficiently tested and still require more thorough characterization. The choice of antibodies cannot be based solely on western blotting. Although it is critical that an antibody bind only the appropriate band(s) in a western, some antibodies may not react at all in western blotting and work fine in IHC assays and vice versa. This is in part because discontinuous epitopes are generally denatured in western blotting while formalin fixation tends to preserve the tertiary structure of proteins and spare discontinuous epitopes. For some targets there are obvious antibody choices since an antibody may be very well characterized and in standard use in commercial diagnostic laboratories. For example, the MIB-1 clone for Ki-67 is the antibody of choice for evaluation of proliferative activity.

IHC testing parameters

When testing a new antibody to a biomarker target, where do you start? Here we introduce some guidelines for initial through final testing. After selecting antibodies to a biomarker one must then consider: which tissues to perform the testing on, the type of tissue pretreatments, if any, the most appropriate detection system, and the optimal antibody concentration.

Tissues in assay development

Appropriate tissues need to be chosen for IHC testing of the antibodies in FFPE tissues. The choice of tissues should include a tissue(s) known or thought to express the target protein. In addition, the target tissue type should be tested to determine the antigen expression level. For example, if the clinical study involves testing samples from patients with an inflammatory disease such as Crohn's disease then several Crohn's tissues should be included in the assay development testing. An important point to consider when selecting tissues for optimization is that there can be sample-to-sample variability of the target protein of interest, thus optimization should include a large enough replicate group of samples to recognize this range. It is also important to choose target tissues that demonstrate an array of expression levels of the target protein from low to high. High expression tissue may be valuable in initial testing to demonstrate which antibody binds well, but care must be taken so that the assay developed is sensitive enough to detect proteins in tissues where the expression is lower. Also, for many targets in many disease types there are variable levels of expression among patient samples. Some tissues may not express the target at all and others may demonstrate very high levels of staining. It thus makes sense to use multiple different tissues other then the target tissues in question so as to gain a confidence level in understanding the spectrum of expression that can be expected in the clinical study specimens. Sometimes more than one tissue type will be tested for IHC biomarker staining and each needs to be tested with the assay prior to the study. For example, in some oncology studies tumor tissues may be tested as well as skin biopsies, which may be used as a surrogate tissue. Bone marrow biopsies and lymph nodes may also be included.

The size of the tissues used in testing is also important. Very small tissues, such as those used in tissue microarrays, may not be appropriate depending on the disease tissue under evaluation. Additionally, small samples may fail to detect heterogeneous expression patterns in tissues. Often in tumor tissue some target proteins are present only in a subset of tumor cells, which may be unevenly distributed throughout the tissues. In addition, targets may be expressed in stroma or sometimes in infiltrating cells as well as the tumor. This holds true for various inflammatory diseases, where the pathologic tissue is interspersed among normal background tissue. Thus, larger tissue samples should be evaluated when assessing IHC staining patterns.

Tissue pretreatments, antibody concentrations and choice of detection systems

Developing quality IHC assays in FFPE tissues usually requires pretreating the tissue sections after the tissue sections are deparaffinized. This pretreatment often requires heating in the presence of various solutions, such as sodium citrate in order to increase the availability of the target epitope to the antibody. This process has been referred to as antigen retrieval or, more appropriately, heat-induced epitope retrieval (HIER) (Gown et al, 1993). Heating may be obtained via steaming, in an autoclave or pressure cooker, or in a microwave oven. A variety of pretreatment solutions are commercially available. Enzyme digestion with various proteolytic enzymes such as pepsin, trypsin and proteinase K may also be used as tissue pretreatments to unmask epitopes blocked by the fixation process in FFPE tissues. The numerous tissue pretreatments available for testing greatly increase the number of slides that need to be initially tested, but this is crucial to the optimization. Different pretreatments often yield very different results and the best pretreatment can only be determined by empirically evaluating a variety of pretreatment procedures. For example, some pretreatments may result in cytoplasmic staining in appropriate cell types for a target that is expected to be membrane associated. Another pretreatment may yield the membrane as well as the cytoplasmic staining. Deciding which pretreatment is optimal depends on a variety of factors, such as what is known about the micro-anatomical and cell type distribution of the target antigen.

The optimal concentration of each antibody for IHC assays is variable; although most fall within the range of 0.1–5.0 μg/mL. The antibody concentration depends on its incubation time, e.g., 30–60 minutes or overnight, the sensitivity of the detection system and the characteristics of the antibody and target protein, i.e., antibody affinity and availability and expression level of its target molecule. In some cases the antibody concentration may fall outside this range; however, if concentrations higher than 10.0 μg/mL are used, they are likely to result in non-specific staining. A range of antibody concentrations should be used in initial testing using the various pretreatments described above. The antibody concentration can then be adjusted and the tissue pretreatments narrowed down to acquire the most sensitive and specific staining that achieves minimal background staining. The final antibody concentration should be chosen after testing a variety of tissues with different expression levels.

A variety of well characterized colorimetric detection systems are commercially available. These range greatly from the source of the label, e.g., biotin, FITC; enzymatic reaction used, e.g., horseradish peroxidase, alkaline phosphatase, etc.; the chromogen, and counterstains. The tissue type being tested in the study may be the most important consideration in dictating the choice of detection system. If the tissue type is rich in endogenous biotin then a non-biotin-based detection system may be necessary. If the tissue type is highly melanotic, a detection system with a chromogen that can be easily distinguished from the melanin may be more useful. In most cases we have found that a peroxidase-based detection system with DAB as chromogen and hematoxylin counterstain provides the best localization and resolution of a target protein. The experience level and preferences of the individuals analyzing the assays is also a strong consideration.

Controls in IHC assays

Running proper controls is critical in understanding IHC reactivity when developing assays. Negative controls should be tested on the same tissue sections as the test antibody. It is imperative to use serum negative controls and not just omit the antibody by replacing it with buffer. For mouse monoclonal antibodies, mouse IgG should be run at the same concentration as the test antibody and the tissue sections should be tested under identical conditions. Isotype controls may also be tested. For rabbit monoclonal and polyclonal antibodies, rabbit IgG should be used as the negative control. Some tissues may bind non-specifically to reagents in the detection system. For example some tissues may be rich in endogenous biotin, which will label when avidin-biotin detection systems are used. Review of the negative controls will determine if the staining is due to the binding of the primary antibody or due to some other source. Positive controls should express the target antigen and when possible should also include negative cell types and moderate levels of expression of the target molecule. Positive control tissues are also helpful in identifying the cell types that may be labeling with an antibody. For example, if a target is thought to be expressed in B-lymphocytes, then staining should be detected primarily in the germinal centers, the normal anatomical compartment of B-lymphocytes, in a section of tonsil or normal lymph node. Here the positive control material not only provides a tissue expressing the target but facilitates the determination of the assay's specificity. Ideal positive and negative controls can be identified during the assay development stage and then selected for use in the clinical study.

Assessing IHC reactivity

IHC data should be assessed for both specificity and sensitivity of staining. Specificity of staining can be assessed by subcellular localization, specific cell and tissue type staining and differential staining. IHC staining can readily discriminate whether a target is expressed in the nucleus, cytoplasm or associated with the plasma membrane as discussed previously. The staining pattern for an antibody to a given target should first be checked for the appropriate expected localization pattern. In addition to subcellular localization, specific cell types may be helpful in understating assay specificity. For example, a target such as CD31 that is expressed in endothelial cells of blood vessels should label with this pattern and appropriate tissues should be tested (see Figure 5.1). The sensitivity of the staining can be gauged by testing a number of tissues with different levels of expression. The assay and antibody titer should be able to detect lower expression proteins but not at the expense of increased background that can confound analysis.

In some cases little is known about the expression of the target protein. In these instances the specificity and consistency of staining patterns may help decide the assay conditions and choice of antibody. Analysis of the negative control slide is very important in the antibody and assay decision. Often what is believed to be a specific staining pattern with an antibody can also be observed in the negative isotype control. Thus, some IHC staining patterns appear more real than others, but may be non-specific since they closely follow changes in background staining. The final result should be a reproducible, robust assay. Good support of specificity is provided when more than one antibody to a different epitope of the same target labels with a similar pattern. This is not always the case, however, as some antibodies do not bind well with their epitope in FFPE tissues or there may be different isoforms of the target that have different subcellular localizations. A good example of this is dUTPase, a nucleotide metabolism protein that creates substrate for thymidylate synthase, which is a target for therapy in several cancer types such as colonic and gastric carcinomas. The dUTPase target was thought to be nuclear based on its function; however, IHC staining yielded very specific granular cytoplasmic staining as well as nuclear staining. Later the enzyme was shown to have a mitochondrial isoform and thus the cytoplasmic staining was true and not artifactual. Importantly, nuclear and cytoplasmic staining was scored separately as the nuclear staining predicted response to therapy (Ladner et al, 2000; Tinkelenberg et al, 2003). In this case the antibody recognized an epitope present on both isoforms (see Figure 5.1).

Cell lines with varying levels of target protein or RNA levels may sometimes be used to help understand the specificity of an antibody.

Formalin fixed cells can be prepared into blocks and then paraffin-embedded to more closely match the tissues being tested. However, assays that work on cell lines may not do so when applied to tissue. This may be due to the differences in how cells are organized in tissues and their extracellular matrix, junctions and cell–cell interaction in comparison to the single cell nature of the cell preparations. In addition, RNA levels may not always correlate with protein levels.

What is a validated assay?

At what point is an IHC assay considered validated? The amount of work that can be performed on each target seems endless, but what in practicality is needed? In a validated IHC assay: the staining pattern should be consistent with what is known about the subcellular and tissue type localization of the target; the staining should be reproducible, i.e., during assay development the final assay should be tested on separate days on the same tissues and demonstrate near identical staining; the assay should be clearly defined along with the reagents used; the assay should be tested in more than one tissue, including the target tissue being used in the clinical study; and the staining should be differential, i.e., if the staining is present in all cells with no variation then the assay is of no value and likely non-specific. In most cases it is prudent to test a large panel of the same target tissue type(s) that will be used in the clinical study so that the expression of the biomarker is understood across a large data set. Lastly, the assay should be able to be measured or scored and the staining performed during assay development should provide a good baseline for interpretation (scoring will be discussed in more detail below). After all of these criteria are met the assay should be documented in the form of a final report detailing all of the pertinent assay conditions and results from the assay development. This information can be written into a study protocol and/or SOP specifically for the IHC testing of the clinical samples. This protocol would then be used to test samples from the clinical study. Note that some assays are better than others and there are limits based on available antibodies and expression level of the target proteins.

Development of biomarker scoring scheme

At the end of assay development and validation the manner in which the IHC assays will be measured should be clear but may require some refining after evaluating the clinical study set. Scoring in IHC is antibody/biomarker dependent and depends upon the staining pattern of each target as well as the tissue type being analyzed. The staining pattern of a marker may best be measured according to the percentage of positive cells of a particular cell type. This is common in oncology studies where often only a subset of tumor cells label. In the clinical pathology laboratory, estrogen and progesterone receptors (ER, PR) are measured as a percentage of nuclear positive breast tumor cells. In some instances a histoscore (H-Score), which includes the percentage of immunoreactive cells and the staining intensity, may be clinically more significant. The scoring scale needs to be defined for each marker. Instead of providing a score based on percentage, a scale of 0–3 may often be used. In general, a score of 0 indicates negative staining, a score of 1 low level staining, 2 moderate level staining and 3 high level of staining. In the case of membrane-based antigens the intensity of the immunoreactivity as well as the distribution of the stain along the cell membrane (complete or incomplete) are essential to proper interpretation. Such is the case for HER2 immunostains.

Scoring schemes need to be developed according to the reactivity pattern for each biomarker as well as the disease in which it is being tested. In some cases more than one score per sample per biomarker may be applicable. A target biomarker protein may be expressed in more than one cell type or with different subcellular localization patterns within the same cell type. For example in oncology studies, the target may be expressed in tumor cells as well as stroma and possibly other cell types (see Figure 5.3, top images). It may be important to measure these separately to provide more specific staining information rather than just lumping all staining into one score. The same holds true for other diseases that may affect more than one component of the tissue. For example, in some skin diseases the epidermis and dermis may be involved in the disease process and a biomarker may be expressed in both. Once again it is best to provide separate scores and provide a measure for both areas of the tissue. The expression of biomarkers in various diseases in most cases is not completely understood and thus providing better data resolution may be critical to potentially identifying meaningful differences within pre- and post-treatment patient samples as well as among subjects.

Testing larger data sets during biomarker assay development provides the best support for developing an optimal scoring scheme. Larger data sets assist in the understanding of the type(s) of staining results that may be obtained in clinical samples as well as the range of staining and penetrance of biomarker expression within a certain disease tissue type. This data should provide further validation of the assay. Results can be compared to other in-house and any published studies regarding the expression of the target. Proper controls are also critical in using a scoring scheme so that an acceptable range of staining can be determined and help determine whether the assay was performed properly and whether the tissues can be scored.

When possible the control tissues should represent a range of staining and should be used throughout the study to monitor any potential protocol drift.

In some studies the use of a second reviewer may be important to read a subset of the samples to ensure the scoring consistency. If more than one reviewer is required, the reviewers need to initially agree on the scoring method and to 'calibrate' the method. This can be done by picking a subset of the samples that demonstrate various staining patterns, having them scored separately, and then have the two reviewers review the slides together and agree on the scoring.

Quantification – image analysis

Some assays may require a quantitative approach that goes beyond the visual microscopic evaluation of a pathologist or immunohistochemist. Our laboratory uses an approach we term 'rational image analysis'. In this approach the basic understanding is that one size does not fit all when performing image analysis of biological tissues. Before attempting an analysis using computer assisted image analysis software, an understanding of the biomarker and its reactivity in the tissues is of chief importance. There are many facets to biomarker staining that need to be addressed. Some important questions include: what should be measured, what cell types should be measured and of these cell types, should a percentage of positive cells be measured or the total number of positive cells be counted? Should intensity of staining be measured and should this be linked to the percentage of positive cells as H-scores? What compartment is to be measured, i.e., nucleus, cytoplasm, membrane or a combination thereof? What portion of the tissue or what cell types are to be quantified? In addition, what is the intra-specimen variation and what is the variation in measurement based on specimens stained or analyzed on different days. After compiling all of this information, a rational approach to the analysis can be determined. It is important to reiterate that scoring schemes and the method of analysis depends on the IHC biomarker staining pattern and the pathology of the tissue type being measured.

In many cases, microscopic computer assisted image analysis may not be the best method of analysis and scoring of the IHC assay. A pathologist or immunohistochemist may better measure the overall staining pattern of the IHC stain by providing a visual microscopic analysis without the use of image analysis software. In most cases, if changes in staining patterns cannot be seen by microscopic examination with the human eye, then using image analysis software will likely yield the same result. Microscopic visual examination may be semiquantitative by having an experienced microscopist estimate the quantity and/or intensity of positive cells or quantitative by performing actual cell counts. Both analyses have their attributes and drawbacks. For instance, the semiquantitative approach has less granularity but is typically the quicker and more cost-effective method when finer resolution is not needed. Quantitative visual microscopy is slow and arduous but can potentially demonstrate differences on a finer resolution. Either method of microscopic visual examination may provide a very reliable score that can be consistently applied to the dataset, providing high quality data. The value of microscopic visual examination is underlined when measuring a biomarker with heterogeneous expression across a sample as well as one that may be present in more than one cell type or cell compartment or in various tissue pathologies. Scoring IHC assays by either computer-assisted analysis or visual microscopic analysis is a complicated task that requires a well thought out practical approach by experienced individuals.

5.4 IHC testing and analysis of clinical specimens using validated assays

This section will focus on the testing and analysis of clinical study tissue specimens. In addition to specimen testing this section also focuses on the subsequent analysis and scoring of the stained samples as well as the management and transfer of data. All facets of IHC testing and analysis need to be quality controlled and well documented in hardcopy with signatures according to the laboratory's SOPs.

The frequency or timing of testing of the study specimens needs to be defined. In some studies the staining is performed after all of the patients have been enrolled in the study and all of the samples collected. Enrollment in studies may range from months to years. For studies that take long to enroll or include ascending dose cohorts, interim analyses may be required. It is important that the assay is consistently performed and analyzed and that there is no drift in the protocol between interim analyses. Thus, before each interim analysis a quality control run should be performed.

Quality control test

Before any IHC assays are run on sections of the clinical study samples, a quality control run should be performed to ensure the assays are working properly and yield equivalent results when compared to previous runs. This should be tested on control tissue(s) identified in the assay development and validation stage of the study. This run should include the primary antibodies, negative controls and detection system reagents that will be used in the testing of the clinical specimens. The staining patterns observed in the control tissues of the quality control test will determine if the assay is working appropriately.

IHC testing of clinical samples

Clinical samples are tested using the validated assay conditions and reagents. To promote consistency, it is recommended that all samples be tested at the same time using the same set of reagents. When all samples are received or when a cut-off date or number of specimens is set for interim analysis, a set number of serial sections from each sample block should be prepared. The first and last section may be stained with hematoxylin and eosin as a reference for understanding tissue histology and pathology. Prior to IHC staining of these samples a quality control run should be performed with the test antibody and the results reviewed by the study director for acceptable staining. If acceptable, proceed to testing the clinical samples. If five different biomarkers require testing, each of the sections should be assigned to the test antibody along with its appropriate negative control. For example, if Ki-67 is being tested, it should be tested on the same number section of every sample and its negative control on a serial section. All other markers should be tested on subsequent sections in a similar fashion. Appropriate positive control tissue(s) should also be run. IHC staining runs should be well documented, including lot numbers and dilution calculations for all reagents used. After staining, all study samples should be securely stored until ready for analysis.

IHC staining results should initially be screened to make certain that all of the tissues stained appropriately. First, the staining results of control tissues should comply with the acceptance criteria of the assay. Each stained section from each clinical sample, both the test antibody as well as the negative control, should be reviewed to determine if the sections are acceptable for analysis. If a particular sample did not stain appropriately, e.g., folds in the tissue, incomplete staining or tissue sections peeling off of the slide, then these samples should be repeated along with their negative controls and the positive control. This work should also be well documented. A complete set of clearly identified stained samples for each IHC biomarker should be submitted for analysis. As with any GLP study, all necessary documentation should be completed according to the laboratory SOPs. A simple rule of thumb for documentation is: does it show that the study was performed as prescribed by the study protocol and laboratory SOPs and is all work traceable?

Analysis and scoring of biomarkers on sections of clinical samples

Once the clinical specimens are stained and deemed acceptable, the samples are submitted for analysis and scoring. The scoring scheme developed during assay development and validation must be consistently applied to the dataset. It is important to be consistent in the application of the scoring criteria to all samples. This may be particularly challenging when studies are performed with several different interim analyses over long periods of time. Proper control samples help calibrate the scoring so that they may be retrospectively referred to in order to make sure there is no drift in the scoring.

Prior to scoring clinical samples, the reviewer should evaluate all of the samples and then begin scoring. All analyses of clinical samples are performed under blinded conditions. When scores are generated they should be entered into a database or spreadsheet in a controlled fashion according to the SOPs of the laboratory. Once the data are completely scored and submitted and approved for quality control, the data need to be sent to the sponsor according to a previously agreed upon and tested format and method of transfer.

Data management and data transfer

The manner in which IHC results and sample data are generated, managed and transferred also needs to be well controlled and documented. The method of scoring will dictate the databases that need to be generated and populated for each sample and biomarker. The IHC laboratory needs to work with the data management group to identify how each sample will be designated and the types and parameters of scores assigned for each biomarker. The final database needs to be built and then test files generated and transferred to make sure that the information is integrated and transferred appropriately. Data security is paramount and should be addressed. Examples of secure transfer include password protected documents that may be sent via e-mail or web-based interface or sent via courier on CD Rom.

In addition to sending the scoring data a final report may also be required. This final report should summarize the IHC method used and an explanation of the scoring criteria. The final report should follow FDA guidelines as contained in 21CFR58.185. These include a quality statement, lot numbers and expiration dates of reagents used, personnel involved in the study and location of all study records and materials. The study sponsor may request that representative or specific digital images are also included in the final report.

5.5 Applications of IHC testing in clinical studies

Immunohistochemistry assays have broad application in clinical studies, particularly for assessing protein biomarkers. IHC biomarker assays may be used to label subject tissue samples with the intent of providing information predictive of patient response to a specific drug therapy or the effectiveness of a drug. These assays have varied applications based on the questions they ask and at what stage of clinical development they are tested. IHC biomarker assays may be placed into a number of different categories, e.g., pharmacodynamic or pharmacodiagnostic depending on how they are assessed in the clinical study. Whatever the case, it is important to develop quality reproducible assays that can be applied to the clinical studies.

The IHC biomarker assays are normally measured in pre- and post-treatment tissue samples to provide an understanding of changes in a biomarker upon therapeutic treatment. Pre-treatment samples may come from earlier diagnostic tissue blocks or in many cases may be taken as baseline samples at the start of the study. Post-treatment samples are obtained after the subject has received the therapy. In some cases more than one post-treatment sample may be included in a study to measure changes over time. IHC biomarker assays are tested on sections of the set of samples and scored for reactivity in a blinded fashion. After unblinding, the results are analyzed to determine any changes in staining levels.

In earlier stages of clinical development IHC assays may be used to determine whether a drug therapy is active based on changes observed in expression of a target protein upon treatment. These are referred to as pharmacodynamic biomarkers. For example, if a therapy is designed to inhibit cell division, then Ki-67, which is present in the nucleus of only non-G_0 cells, may be a good IHC biomarker to determine whether the therapy is having its desired effect, i.e., there should be an overall decrease in the number of Ki-67 positive cells in the post-treatment sample. Similarly, if the same therapy promotes apoptosis, then assays of apoptotic biomarkers would be expected to increase if the therapy is effective. These markers may have value in the disease target tissues, e.g., tumor samples, as well as surrogate tissues, which may be more readily available to measure (Albanell et al, 2002). The IHC assays may also be used in studies where multiple ascending doses are used. These assays may help in choosing dosing levels of the drug based on objective changes observed in the pre- and post-treatment samples at various dosage levels.

In later stage clinical studies the IHC biomarker assays may be used to help stratify patient populations to determine which patients may respond to therapy and therefore be eligible for enrollment. These assays are particularly relevant when evaluating targeted therapies where it is important to understand whether a patient is expressing a particular protein to which the therapeutic is directed. These types of assays are referred to as predictive or pharmacodiagnostic tests. The HercepTest™ is an example of an IHC test that is used to determine whether a sample demonstrates elevated levels of HER2 protein and thus is a candidate for Herceptin™ therapy (Dowsett et al, 2003; Garcia-Caballero et al, 2006). Other IHC biomarkers may provide patient response information to a therapy based on outcome data, but these markers are not the actual targets of the therapy. These biomarkers are referred to as surrogate markers. Favorable and unfavorable prognostic markers may be identified and may be useful in future studies to potentially stratify patients by helping to choose candidates for studies that will most likely respond and thus increase patient response rate.

The FDA has made a recent push for pharmaceutical and biotechnology companies to include the use of biomarkers in drug development and testing and has provided some guidance to the industry (CDER Guidance for Industry – Bioanalytical Method Validation, May 2001). As a result, IHC biomarkers are currently being evaluated in numerous clinical studies. Our group is currently involved in many different clinical studies that involve IHC biomarker analysis. We have demonstrated successful IHC biomarkers in oncology as well as immune-related diseases. Examples from two different studies are described below showing broad applications of IHC assays in GLP clinical studies.

Our first example demonstrates IHC staining of tissue samples from a clinical study in subjects with psoriasis. At baseline, in the untreated psoriatic sample, CD4 levels are extremely elevated in the dermis and the epidermis is thickened compared to normal skin. CD4 is a biomarker that labels a subpopulation of T-lymphocytes that are shown to dramatically decrease upon drug treatment. Upon treatment a decrease in CD4 positive T-cells is also accompanied by a thinning of the epidermis (Figure 5.2). Both the high level of T-cells and thickened epidermis are hallmarks of psoriasis and both are shown to change dramatically upon drug treatment to be consistent with what is expected in normal skin samples. This is the expected outcome upon therapeutic intervention.

Another example is from a study using a targeted drug therapy in oncology where IHC is used not only to detect the presence of the target protein of the drug, but also to detect the presence of the monoclonal

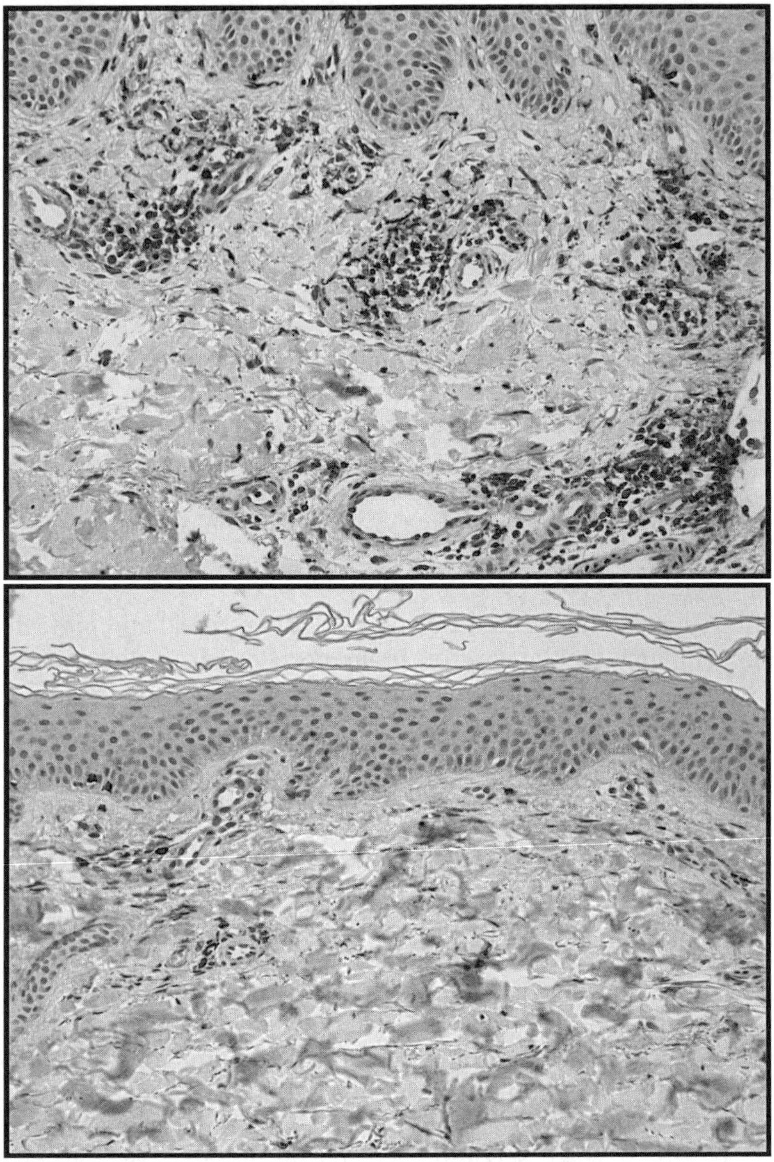

Figure 5.2 Baseline and post-treatment skin sample biopsies from a patient with psoriasis. **Top:** Baseline skin sample biopsy from a psoriatic patient. There is an abundance of CD4 positive cells in the dermis of the tissue. This tissue has a thickened epidermis (not shown). **Bottom:** Post-treatment skin sample biopsy. There is a great diminution in CD4 positive cells in the dermis along with a thinning of the epidermis. CD4 content and epidermal thickness is more in-line with normal skin demonstrating the effect of the therapy. Positive staining is indicated by the presence of the dark brown chromogen. Hematoxylin counterstain. A full-colour version of this image appears in the colour plate section of this book.

antibody drug in the post-treatment tissue. IHC can be used to determine the levels of expression of the protein target as well as the cell types that are expressing the target (Figure 5.3, top images). IHC can also be used to detect the presence of an antibody therapeutic in a tissue, given that an IHC-reactive anti-idiotype antibody reactive to the therapeutic antibody is available (Figure 5.3, bottom images). These IHC biomarker assays may help predict which patients may be eligible for therapy based on the IHC staining pattern and may also help determine if the drug is reaching the tumor cells.

IHC biomarker assays may not only prove to have a positive impact on current clinical studies but some may have the potential for further

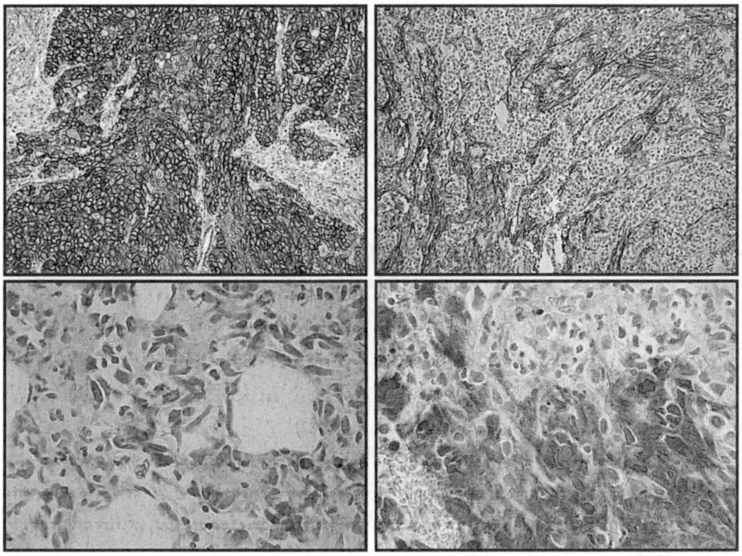

Figure 5.3 Variable expression of target for targeted therapy and anti-idiotype immunohistochemistry. **Top: variable IHC staining of biomarker target for targeted therapy. Top Left:** Esophageal carcinoma showing strong plasma membrane staining of tumor cells and minor staining of stroma with an antibody directed against the target protein of a targeted antibody therapeutic. **Top Right:** Esophageal Carcinoma: Strong stromal staining and no tumor staining for the same target shown at left is shown in this tumor from a different patient. **Bottom:** Anti-idiotype immunohistochemistry to detect localization of antibody therapeutic. **Bottom Left:** Baseline tumor sample prior to antibody therapy. As expected, no specific staining is detected. **Bottom Right:** Post-treatment sample demonstrating strong anti-idiotype staining. The anti-idiotype antibody is directed against the antibody therapeutic demonstrating that the therapeutic antibody is localized in the tumor. Positive staining is indicated by the presence of the dark brown chromogen. Hematoxylin counterstain. A full-colour version of this image appears in the colour plate section of this book.

development resulting in new clinical tests to be used in the hospital pathology setting as diagnostic or prognostic tests to monitor and guide patient treatment.

References

Albanell J, Rojo F, Averbuch S, Feyereislova A, Mascaro JM, Herbst R, LoRusso P, Rischin D, Sauleda S, Gee J, Nicholson RI and Baselga J (2002). Pharmacodynamic studies of the epidermal growth factor receptor inhibitor ZD1839 in skin from cancer patients: histopathologic and molecular consequences of receptor inhibition. *J Clin Onc*, **20**, 110–124.

ASCO (1996). Clinical practice guidelines for use of tumor markers in breast and colorectal cancer. *J Clin Oncol*, **14**, 2843–2877.

Battifora H (1991). Assessment of antigen damage in immunohistochemistry. The vimentin internal control. *Am J Clin Pathol*, **96**, 669–671.

CDER Guidance for Industry – Bioanalytical Method Validation, May (2001). Guidance document prepared by the Biopharmaceutics Coordinating Committee in the Center for Drug Evaluation and Research (CDER) in cooperation with the Center for Veterinary Medicine (CVM) at the Food and Drug Administration.

Darnell Jr JE (1997). STATs and gene regulation. *Science*, **277**, 1630–1635.

Dowsett M, Bartlett J, Ellis IO et al (2003). Correlation between immunohistochemistry (HercepTest) and fluorescence in situ hybridization (FISH) for HER-2 in 426 breast carcinomas from 37 centres. *J Pathol*, **199**, 418–423.

Garcia-Caballero T, Menendez MD, Vazquez-Boquete A, Gallego R, Forteza J and Fraga M (2006). HER-2 status determination in breast carcinomas. A practical approach. *Histol Histopathol*, **21**, 227–236.

Gown AM, De Wever N and Battifora H (1993). Microwave-based antigenic unmasking: A revolutionary new technique for routine immunohistochemistry. *Appl Immunohistochem*, **1**, 256–266.

Ladner RD, Lynch FJ, Groshen S, Xiong YP, Sherrod A, Caradonna SJ, Stoehlmacher J and Lenz HJ (2000). dUTP nucleotidohydrolase isoform expression in normal and neoplastic tissues: association with survival and response to 5-fluoruracil in colorectal cancer. *Cancer Res*, **60**, 3493–3503.

Riera J, Simpson JF, Tamayo R and Battifora H (1999). Use of cultured cells as a control for quantitative immunocytochemical analysis of estrogen receptor in breast cancer. The Quicgel method. *Am J Clin Pathol*, **111**, 329–335.

Tinkelenberg BA, Lynch FJ and Ladner RD (2003). Identification of sequence determinants of human nuclear dUTPase isoform localization. *Exp Cell Res*, **287**, 39–46.

6 Flow cytometric cell-based assays: an overview of general applications

Cuc Davis, Manjula Reddy*, Thomas Williams and Uma Prabhakar

Centocor Research and Development, Inc., 145 King of Prussia Road, Radnor, PA 19087, USA

6.1 Introduction

Flow cytometry can provide rapid and accurate multivariate measurements of cellular subpopulations from a heterogeneous population of cells to aid evaluation and understanding of immune responses. This application is particularly useful in the clinical setting when assessing the safety and potential efficacy of a therapeutic from a small sample size, mainly whole blood samples from immune compromised patients but can also include bone marrow, serous cavity fluids, cerebrospinal fluid, urine, and solid tissues. Two flow cytometric cell-based assays routinely used in the clinical setting

* To whom correspondence should be addressed.

Validation of Cell-Based Assays in the GLP Setting: A Practical Guide Editors, Uma Prabhakar and Marian Kelley

are immunophenotyping and cell cycle analysis (Brown and Wittwer, 2000). In addition, there are other flow cytometric-based assays for evaluating apoptosis, intracellular cytokine staining, and intracellular signaling by way of phosphorylation, that are becoming increasingly useful in the clinical environment. In this chapter we will review some of these applications and also provide the researcher with critical parameters that must be addressed, as these applications are being developed to support clinical trials.

6.2 Immunophenotyping

Immunophenotyping addresses the identification and quantification of unique cellular antigens on white blood cells using specific monoclonal antibodies directly or indirectly conjugated to a fluorochrome. Key information concerning cell lineage, activation status, adhesion, migration and homing capacity, and the readiness to launch a response to stimuli and interaction with other cells can be inferred by immunophenotyping cell surface molecules involved in mediating immune responses. Not only is the technique extremely specific, the assays are quick with the ability to analyze a large number of patient samples, and is highly reproducible.

Immunophenotyping is routinely used to monitor human immunodeficiency virus (HIV)/acquired immunodeficiency syndrome (AIDS) patients, to diagnose and classify leukemias, chronic lymphoproliferative diseases, and malignant lymphomas, to monitor residual disease, to monitor progression of patients after chemotherapy and organ transplants, to monitor cells of the immune system to determine safety, and to measure cell surface markers associated with the mechanism of action of therapeutics to determine efficacy.

T lymphocytes play a major role in autoimmune diseases (such as rheumatoid arthritis, psoriasis, and multiple sclerosis), allergic diseases (asthma), and oncologic disorders (Takashima and Morita, 1991 and Owens et al, 2000), as well as other immune mediated inflammatory diseases. For example, changes in T lymphocyte sub-populations and T cell function in T helper 1 or T helper 2 in cell mediated diseases during and after therapy is monitored to evaluate the effect of therapy (Rose et al, 1985 and Rogge, 1997).

Early activation markers (CD69) and late activation markers (CD25 and HLA-DR) are evaluated in many clinical studies (Masatoshi et al, 1997; Ferenczi et al, 2000, and Asadullah et al, 1997). Increasing circulating HLA-DR+ and CD25+ T lymphocytes have been reported in patients with active psoriasis (Paciel et al, 1984 and Ferenczi et al, 2000); CD25 (interleukin-2 receptor alpha chain, IL-2R) has been widely studied as the activation marker to evaluate systemic T-cell activation in many clinical studies (Khoury et al, 2000; Ferrarini et al, 1998; Strauss et al, 1995; Lemster et al, 1994, and Mantovani et al, 1994); CD25 antigen has been detected at a low density on normal T and B cells. Increased levels of peripheral HLA-DR+ T lymphocytes have been reported in immunopathogenic diseases and after immunostimulation (Tomkinson et al, 1987, Masatoshi et al, 1997, Ferenczi et al, 2000, and Asadullah et al, 1997). Increased expression of HLA class II antigens (DR) on peripheral blood helper T cells (CD4+HLA-DR+) and CD4+IL-2R+ (CD4+CD25+) were found in relapsing MS patients, suggesting a systemic T-cell activation in multiple sclerosis (MS) (Scolozzi et al, 1992).

Chemokine receptors (CKR), a family of structurally related seven-transmembrane G protein-coupled receptors (CXCR and CCR), involved in regulating numerous T lymphocyte functions, most importantly, regulating the recruitment of leukocytes in inflammatory conditions (Hancock et al, 1996) are differentially expressed on memory T lymphocytes depending on their polarization, with Th1 lymphocytes expressing CXCR3 (Qin et al, 1998). The accumulation of CXCR3+ cells at the sites of inflammation of the Th1-mediated diseases, rheumatoid arthritis, and multiple sclerosis (Calabresi et al, 2002, and Kivisakk et al, 2002) has been reported. It was shown that CXCR3+ T lymphocytes levels increased in blood of relapsing-remitting and progressive MS (Balashov et al, 1999). Furthermore, CXCR3 is implicated in the pathogenesis of psoriasis and has been identified to be the key receptor that mediates T lymphocyte trafficking to the overlying epidermis (Rottman et al, 2001).

Immunophenotyping is also used to understand the cellular basis of immunological memory, e.g., T cell function in the pathogenesis of autoimmune diseases is also important. Changes in the expression of markers associated with naïve and memory phenotype have been reported in diseases involving immune activation (Porrini et al, 1992, Lemster et al, 1994, Crucian et al, 1995, Strauss et al, 1995, Abe et al, 1997, and Natelson et al, 1998). In humans, differentially spliced isoforms of the leukocyte common antigen (CD45) have been used to distinguish naïve and memory cells. The highest molecular weight isoform of CD45 (a cell surface glycoprotein of 200 KD complex that is selectively expressed on almost all hematopoietic cells), termed CD45RA, is expressed on a stable population of cells that divide approximately once in every two years. These CD45RA+ cells respond poorly to recall antigens and when activated lose expression of the CD45RA isoform, but instead gain the expression of the low molecular weight isoform, CD45RO. The CD45RO+ population responds rapidly to recall antigens, and they divide about once every 2 weeks *in vivo* (Akbar et al, 1988, and Faint et al, 2001).

A number of critical factors affecting the reliability of the assay, including specimen collection, specimen transport, sample processing, and data analysis, to name a few, must be monitored during the development of this assay. These factors are reviewed in the following sections.

6.3 Critical parameters to be evaluated for immunophenotyping assays

Specimen collection

Specimen collection is the first aspect that can affect the reliability of a flow cytometric assay for immunophenotyping. The collection of whole blood requires knowledge of the purpose of the experiment and proper technique in drawing the blood, such as using the proper needle gauge, appropriate site of collection, and discarding the first draw in the case of immunophenotyping platelets. Whole blood must be collected into a venipuncture-evacuated tube or suitable vessel containing an appropriate anticoagulant, preferably tripotassium ethylenediamine tetra-acetate (K_3EDTA) or heparin. It has been noted that if there is a need to store whole blood for longer than 20–30 hours before analysis can begin, K_3EDTA should not be used due to changes in the proportion of lymphocyte subsets compared to fresh blood (Nicholson et al, 1993), and the lack of antigen-specific responses from the removal of intracellular calcium and metal ions (Kumar and Satchidanandam, 2000). However, each cell surface marker should be tested in various anticoagulants over a time-course to elucidate any differences. Acid citrate dextrose is not a recommended anticoagulant for the determination of absolute cell counts, due to its liquid state in the venipuncture tube, which makes it difficult to determine the exact sample volume. The sample volume is important when using a single platform method to enumerate absolute cell counts. The whole blood should be mixed thoroughly with the anticoagulant to avoid clotting. The tube must be labeled with a confidential unique patient identifier, collection date and time, and any other relevant information such as expiration date.

Specimen stability and transport

The reproducibility and quality of data from such assays depend largely upon the stability of the specimen, which is influenced by several factors including transport from the site of collection to the laboratory where it will be analyzed. It has been shown that cell surface expression of many cellular antigens decreases over time when whole blood is stored at room temperature (Ekong et al, 1993), and that the most noteworthy factor affecting the stability is temperature (Paxton and Bendele, 1993). Severe temperatures affect the integrity of the sample. For example, extremely hot or cold temperatures can cause apoptosis, hemolysis, or freeze the blood. Whole blood should be shipped at room temperature unless validated otherwise.

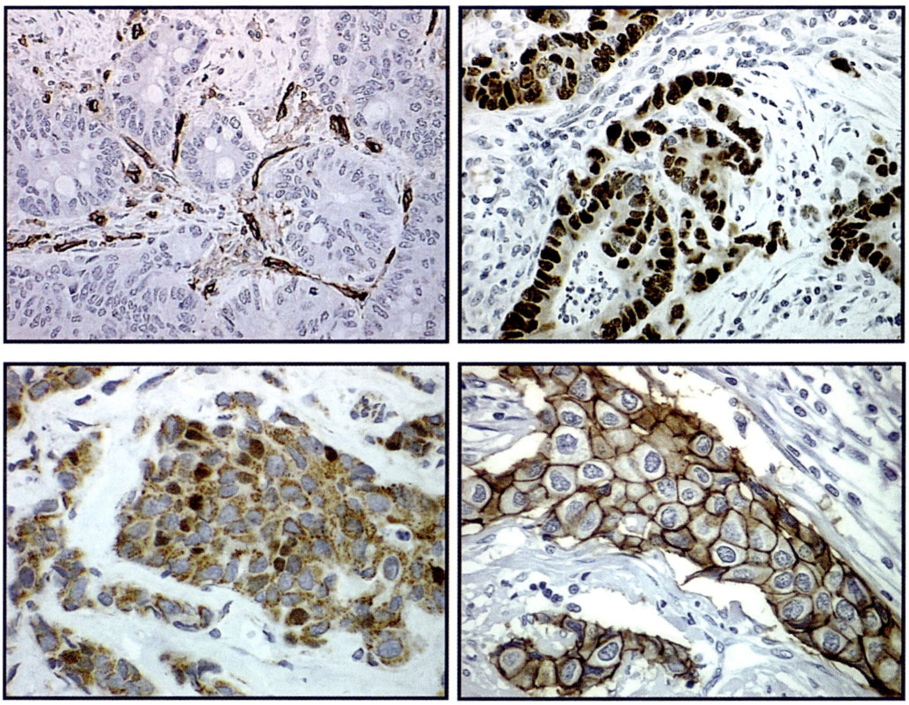

Plate 1 Various examples of IHC staining. **Top Left:** CD31 IHC in a colon carcinoma. The CD31 antibody labels endothelial cells of blood vessels. Tumor cells do not label. **Top Right:** p53 IHC in a colon carcinoma showing strong nuclear localization in tumor cells. **Bottom Left:** duTPase IHC in a breast carcinoma showing both cytoplasmic (mitochondrial) and nuclear localization in tumor cells. **Bottom Right:** HER2 IHC in a breast carcinoma showing strong plasma membrane localization of tumor cells. Positive staining is indicated by the presence of the dark brown chromogen. Hematoxylin counterstain. (See Figure 5.1).

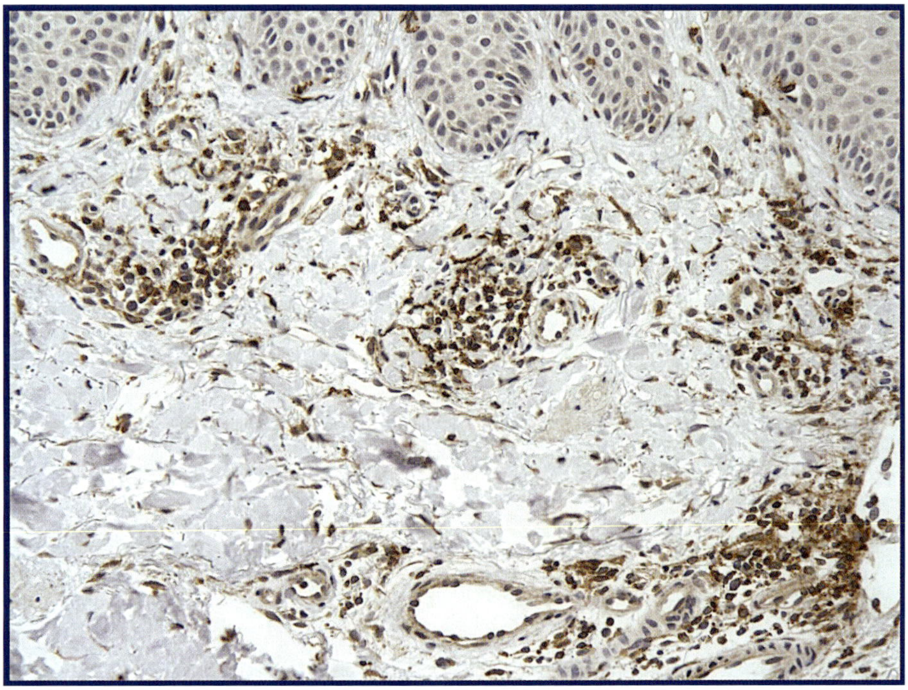

Plate 2 Baseline and post-treatment skin sample biopsies from a patient with psoriasis. Baseline skin sample biopsy from a psoriatic patient. There is an abundance of CD4 positive cells in the dermis of the tissue. This tissue has a thickened epidermis (not shown). (See Figure 5.2).

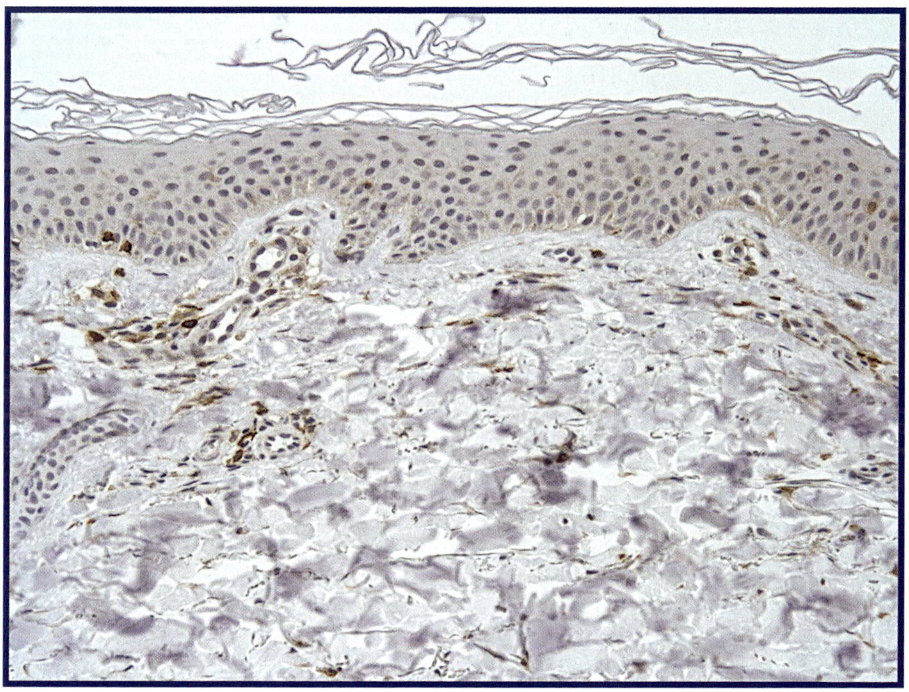

Plate 3 Baseline and post-treatment skin sample biopsies from a patient with psoriasis. Post-treatment skin sample biopsy. There is a great diminution in CD4 positive cells in the dermis along with a thinning of the epidermis. CD4 content and epidermal thickness is more in-line with normal skin demonstrating the effect of the therapy. Positive staining is indicated by the presence of the dark brown chromogen. Hematoxylin counterstain. (See Figure 5.2).

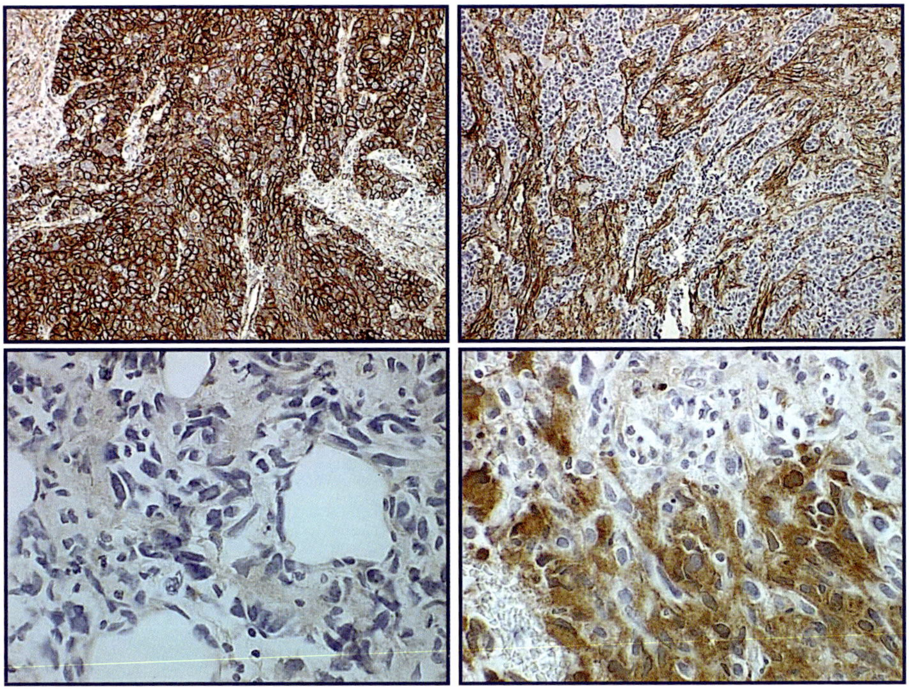

Plate 4 Variable expression of target for targeted therapy and anti-idiotype immunohistochemistry. **Top: variable IHC staining of biomarker target for targeted therapy. Top Left:** Esophageal carcinoma showing strong plasma membrane staining of tumor cells and minor staining of stroma with an antibody directed against the target protein of a targeted antibody therapeutic. **Top Right:** Esophageal Carcinoma: Strong stromal staining and no tumor staining for the same target shown at left is shown in this tumor from a different patient. **Bottom:** Anti-idiotype immunohistochemistry to detect localization of antibody therapeutic. **Bottom Left:** Baseline tumor sample prior to antibody therapy. As expected, no specific staining is detected. **Bottom Right:** Post-treatment sample demonstrating strong anti-idiotype staining. The anti-idiotype antibody is directed against the antibody therapeutic demonstrating that the therapeutic antibody is localized in the tumor. Positive staining is indicated by the presence of the dark brown chromogen. Hematoxylin counterstain. (See Figure 5.3).

Whole blood samples should be tested over a period mimicking the anticipated receipt time of clinical samples to establish maximum storage time before analysis. Venipuncture tubes containing a stabilizer are commercially available and should be used. Validation of fixative tubes (Cyto-Chex BCT from Streck Laboratories) must demonstrate comparative scatter and fluorescence profiles to blood collected in tubes containing anticoagulants and length of storage. During validation, blood samples should be stored or shipped at room temperature and with ice packs to determine the temperature at which the expression of the interested cell surface markers is most stable. Once the samples have been received, sample integrity should be noted, i.e. hot or cold to the touch. If the blood sample feels too cold or too warm, do not attempt to quickly bring the sample to room temperature; Paxton and Bendele (1993) illustrated this might result in spurious data. If there is a visible clot in the blood, the sample is to be rejected.

Specimen staining

The method for staining cell surface markers on whole blood is simple and can be semi-automated. Blood used in the assay should be within the pre-determined time frame determined from stability assessments. A positive control, such as a positive phenotyping panel, must be included during testing to ensure proper sample collection, transport, and processing. The need for FcR block before the addition of monoclonal antibodies against the receptor of interest should be tested during the optimization stages of assay development. Specificity of the fluorescently tagged antibody should be validated by comparing single vs. multiple reagents. The samples should be stained at the manufacturers' and previously titrated concentrations of specific monoclonal antibodies and then incubated with whole blood at room temperature for 30 minutes in the dark. Red blood cells in the stained sample are lysed while the unbound fluorescently tagged antibodies are removed by washing samples in staining buffer followed by fixation if acquisition was to be performed at a later time (usually within 24–96 hours). This step can be automated using a Lyse/Wash Assistant from Becton Dickinson. The cells are then stored in the dark at 4°C until acquisition on the flow cytometer. Prolonged exposure to fixative can be deleterious to the cells; hence the maximum length of fixation should be tested during optimization. Acquisition of the cells in 5 mL polystyrene tubes or 96-well plates can also be automated. Before acquisition of the samples, the flow cytometer must be calibrated daily using beads conjugated to fluorochromes with known light and fluorescence scatter properties. An electronic graphical report, Levey-Jennings, is generated and stored to track the instrument's alignment and performance. The summary reports must be printed and saved in the instrument's daily use logbook.

Data analysis

Standardization of data analysis contributes significantly to the reproducibility of the assay. The gating of lymphocytes should ensure that only the population of interest is included. Quadrant markers are set based on the isotype controls from the same manufacturer as the reagent used to stain the receptors of interest. Quadrant markers should remain constant between replicates, time-points, and patients unless warranted by an abnormal forward vs. side scatter profile, and then logical gating is instituted.

Lymphocyte purity should be at least 85% but optimally be equal to or greater than 95% (MMWR 1997). This can only be checked if cells are also stained with CD14 and CD45 or stained with antibodies against T, B and NK receptors. The lymphocyte purity can be determined by adding the percent positive values for CD3+, CD19+, and CD16+CD56+ cells with a standard deviation of 5% (MMWR 1997). Furthermore, percent positive values for CD3+CD4+ and CD3+CD8+ should add up and equal the percent positive value for total CD3+ with a standard deviation of 10% (MMWR 1997). To determine the absolute cell numbers of CD4+ and CD8+ T-cells, most laboratories use a three- to four-color antibody cocktail which includes CD45 as one of the antibodies (CD3/CD4/CD45, CD3/CD8/CD45, CD3/CD19/CD45, CD3/CD4/CD8/CD45) to aid in identifying the true lymphocyte population from debris, dead cells, and nonleukocytes.

Cell cycle analysis

Defects in cell cycle regulation are a characteristic feature of tumor cells and mutations in the genes involved in controlling the cell cycle are extremely common in cancer. Proteins in the cytoplasm such as cyclins and cyclin-dependent kinases, which are targets for anti-cancer therapies, control the cell cycle.

Standard flow cytometry methods measure changes in DNA content by staining the nuclei of fixed cells with fluorescent dye. The most commonly used DNA dye is propidium iodide (PI), which can be used to stain whole cells or isolated nuclei. The PI intercalates into the major groove of double-stranded DNA and produces a highly fluorescent adduct that can be excited at 488 nm with a broad emission centered around 600 nm. Since PI can also bind to double-stranded RNA, it is necessary to treat the cells with RNase for optimal DNA resolution (Krishan, 1975). The excitation of PI at 488 nm facilitates its use on the bench top cytometers. The cells are into G1, S and G2/M populations according to total fluorescent intensity (Figure 6.1).

Cell cycle analysis and DNA content measurements are used in the clinic to determine DNA ploidy and S-phase fractions as prognostic markers for node-negative breast cancer (Bagwell et al, 2001). Clinical studies have used

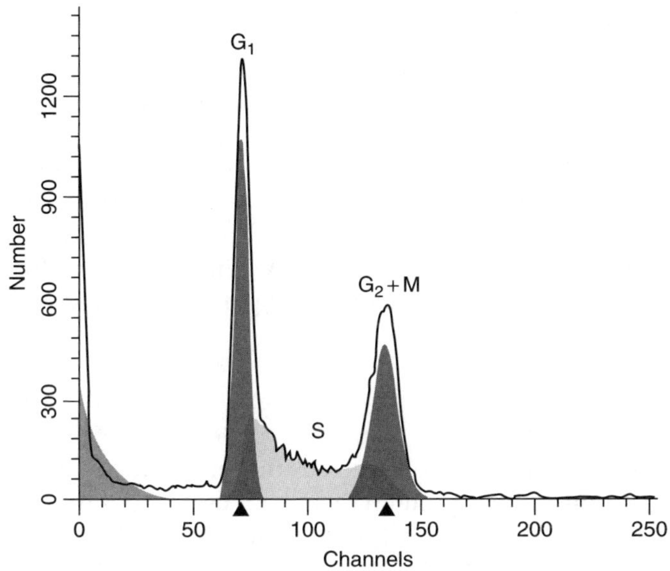

Figure 6.1 Example of analysis of cell cycle histogram

DNA content as an indicator of malignancy found in leukemias, lymphomas, myelomas, and cancer (Barlogie et al, 1983; Brown and Wittwer, 2000).

Apoptosis

Apoptosis can be characterized by the compression of the cytoplasm and nucleus, DNA fragmentation, and the shrinkage of the cell with apoptotic bodies shedding from the irregular plasma membrane exposing phospholipids such as phosphatidlyserine (PS) (Fadok et al, 1992). Two most common flow cytometric assays to detect apoptosis in cells exploit the Annexin V, and the TUNEL assays. Annexin V assay detects the earlier stage of apoptosis by binding to PS in the outer membrane of the dying cell. Also, a positive control for staining should be used; an easily stimulated cell line into apoptosis. The TUNEL assay also known as APO-BRDU™, APO-DIRECT™, or more commonly, "TUNEL" (terminal deoxynucleotidyltransferase nick end labeling) detects apoptotic cells in the later stage of apoptosis, when the DNA undergoes fragmentation. This assay detects fragmented DNA in the cells by using the terminal deoxynucleotidyl transferase enzyme (TdT), which attaches bromolated deoxyuridine triphosphates (Br-dUTP) to the exposed 3'-hydroxyl ends of the DNA. A fluorescent-labeled anti-BrdU monoclonal antibody is used to bind to the Br-dUTP attached to breaks in the DNA (BD Biosciences, 1998).

Intracellular cytokine staining

Cytokines often circulate as proteins bound to soluble receptors, carrier proteins, or inhibitors, which may mask their easy detection by enzyme-linked immunosorbent assay or radioimmunoassay (Dugue et al, 1996, Wadwha and Thotpe, 1998). Many cytokines are undetectable in serum because they are produced locally and have a very short half-life. In order to circumvent these significant limitations, cytokine production is often measured in affected tissues or PBMCs. A general schema for measuring intracellular cytokines by flow is shown in Figure 6.2. This assay is described elsewhere in this book.

Phospho-specific flow cytometry

Phosphorylation is usually a short-lived event related to the activation status of signaling proteins. Flow cytometry can be a useful tool to help determine the correlation between stimuli (growth factors, cytokines, etc.) and the resulting signaling cascade to the phosphorylation state of proteins (Krutzik and Nolan, 2003). Cancers and immune disorders can be phenotyped by comparing differences in signaling events in cells from healthy donors versus

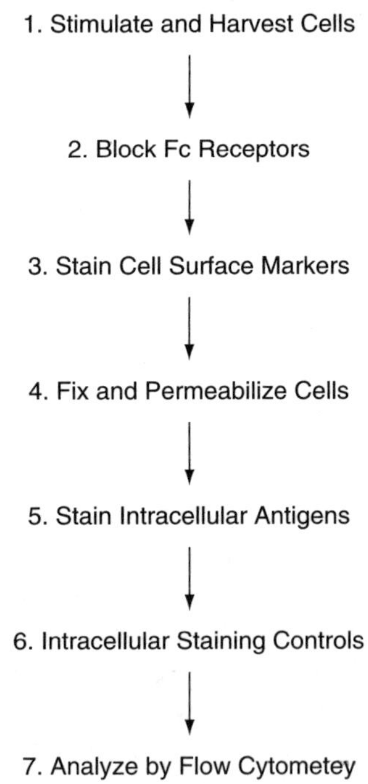

Figure 6.2 Intracellular flow cytometry – overview of staining protocol

cells from donors with specific diseases. Phospho-specific flow cytometry allows for the monitoring of signaling events occurring in distinct subsets of cells. This can be utilized for drug screening and the pharmacodynamic monitoring of patients in clinical trials (Krutzik et al, 2004). This procedure allows for results to be obtained within a day of sample collection.

References

Abe M, Ishikawa O and Miyachi Y (1997). Changes in peripheral blood lymphocyte subsets during cyclosporin administration in patients with psoriasis vulgaris. *Eur J Dermato*, **7**, 417–420.

Akbar AN, Terry L, Timms A, Beverley P and Janossy G. Loss of CD45R and gain of UCHL1 reactivity is a feature of primed T cells. *J Immunol.* 1988; 140 : 2171–2178.

Asadullah K, Friedrich M, Docke WD, Jahn S, Volk HD and Sterry W (1997). Enhanced expression of T-cell activation and natural killer cell antigens indicates systemic anti-tumor response in early primary cutaneous T-cell lymphoma. *J Invest* Dermatol, **108**, 743.

Bagwell CB, Clark GM, Spyratos F, Chassevent A, Bendahl PO, Stal O, Killander D, Jourdan ML, Romain S, Hunsberger B and Baldetorp B (2001). *Cytometry (Commun Clin Cytometry)*, **46**, 121–135.

Balashov KE, Rottman JB, Weiner HL and Hancock WW (1999). CCR5(+) and CXCR3 (+) T cells are increased in multiple sclerosis and their ligands MIP-1alpha and IP-10 are expressed in demyelinating brain lesions. *Proc Natl Acad Sci USA*, **96**, 6873–6878.

Barlogie B, Raber M, Schumann J, Johnson TS, Drewinko B, Swartzendruber D, Gohde W, Andreeff M and Freireich E (1983). Flow ctyometry in clinical cancer research. *Cancer Res*, **43**, 3982–3997.

BD Biosciences (1998). *Apoptosis. Applied Reagents and Technologies Instruction Manual.* 2nd edition, pp 44–65.

Brown M and Wittwer C (2000). Flow cytometry: principles and clinical application in hematology. *Clin Chem*, **46**, 1221–1229.

Calabresi AP, Sung HY, Rameeza A and Whartenby AK (2002). Chemokine receptor expression on MBP-reactive T cells: CXCR6 is a marker of IFNγ-producing effector cells. *J Neuroimmunology*, **127**, 96–105.

Centers for Disease Control and Prevention (1997). Revised guidelines for performing CD4+ T-cell determinations in persons infected with human immunodeficiency virus (HIV). MMWR, 46 (No. RR-2).

Crucian B, Dunne P, Friedman H, Ragsdale R, Pross S and Widen R (1995). Alterations in levels of CD28–/CD8+ suppressor cell precursor and CD45RO+/CD4+ memory T lymphocytes in the peripheral blood of multiple sclerosis patients. *Clinical and Diagnostic Laboratory Immunology*, 2(2): 249–252.

Dugue B, Leppanen E and Grasbeck R (1996). Preanalytical factors and the measurement of cytokines in human subjects. *Int J Clin Lab Res*, **26**, 99–105.

Ekong T, Kupek E, Hill A, Clark C, Davies A and Pinching A (1993). Technical influences on immunophenotyping by flow cytometry: the effect of time and temperature of storage on the viability of lymphocyte subsets. *J Immunol Methods*, **164**, 263–273.

Faint MJ, Annels EN, Curnow SJ, Shields P, Pilling D, Hislop DA, Wu L, Akbar NA, Buckley DC, Moss AHP, Adams HD, Rickinson BA and Salmon M (2001). Memory T cells constitute a subset of the human CD8+CD45RA+ pool with distinct phenotypic and migratory characteristics. *J Immunol*, **167**, 212–220.

Fadok V, Voelker D, Campbell P, Cohen J, Bratton D and Henson P (1992). Exposure of phosphatidylserine on the surface of apoptotic lymphocytes triggers specific recognition and removal by macrophages. *J Immunol*, **148**, 2207–2216.

Ferenczi K, Burak L, Pope M, Krueger JG and Austin LM (2000). CD69, HLA-DR and the IL-2R identify persistently activated T cells in psoriasis vulgaris lesional skin:blood and skin comparisions by flow cytometry. *J Autoimmunity*, **14**, 63–78.

Ferrarini AM, Sivieri S, Buttarello M, Facchinetti A, Perini P and Gallo P (1998). Time course analysis of CD25 and HLA-DR expression on lymphocytes in interferon-β1b-treated multiple sclerosis patients. *Multiple Sclerosis*, **4**, 174–177.

Hancock WW (1996). Commentary: Chemokines and the pathogenesis of T cell-dependent immune responses. *Am J Pathol*, **148**, 681.

Khoury JS, Guttmann RGC, Orav EJ, Kikinis R, Jolesz AF and Weiner LH (2000). Changes in activated T cells in the blood correlate with disease activity in multiple sclerosis. *Arch Neurol*, **57**, 1183–1189.

Kivisakk P, Trebst C, Liu Z, Tucky BH, Sorensen TL, Rudick RA, Mack M and Ransohoff RM (2002). T-cells in the cerebrospinal fluid express a similar repertoire of inflammatory chemokine receptors in the absence or presence of CNS inflammation:implications for CNS trafficking. *Clin Exp Immunol*, **129**, 510–518.

Krishan A (1975). Rapid flow cytofluorometric analysis of cell cycle by propidium iodide staining. *J Cell Biol*, **66**, 188–193.

Krutzik PO, Irish JM, Nolan GP and Perez OD (2004). Analysis of protein phosphorylation and cellular signaling events by flow cytometry: techniques and clinical applications. *Clin Immunol*, **110**, 206–221, 15047199.

Krutzik PO and Nolan GP (2003). Intracellular phospho-protein staining techniques for flow cytometry: Monitoring single cell signaling events. *Cytometry*, **55A**, 61–70, 14505311.

Kumar P and Satchidanandam V (2000). Ethyleneglycol-bis (β-aminoethylether) tetraacetate as a blood anticoagulant: preservation of antigen-presenting cell function and antigen-specific proliferative response of peripheral blood mononuclear cells from stored blood. *Clin Diagnostic Lab Immunology*, **4**, 578–583.

Lemster B, Huang LL, Irish W, Woo J, Carroll PB, Abu-Elmagd K, Rilo HR, Johnson N, Russell-Hall R and Fung JJ (1994). Influence of FK 506 (tacrolimus) on circulating CD4+ T cells expressing CD25 and CD45RA antigens in 19 patients with chronic progressive multiple sclerosis participating in an open label drug safety trial. *Autoimmunity*, **19**, 89–98.

Mantovani G, Maccio A, Lai P, Turnu E and Del Giacco GS (1994). Study of peripheral blood lymphocyte subset distribution and IL-2 receptor (IL-2 R) p55–p75 subunit expression in patients with cancer of different sites. *Cell Biophys*, **24–25**, 301–305.

Masatoshi A, Ishikawa O and Miyachi Y (1997). Changes in peripheral blood lymphocyte subsets during cyclosporin administration in patients with psoriasis vulgaris. *Eur J Dermatology*, **7**, 417–420.

Natelson BH, LaManca JJ, Deny TN, Vladutiu A, Oleske J, Hill N, Bergen MT, Korn L and Hay J (1998). Immunologic parameters in chronic fatigue syndrome, major depression, and multiple sclerosis. *Am J Med*, **105**, 43S–49S.

Nicholson JK, Green TA and Collaborating Laboratories (1993). Selection of anticoagulants for lymphocyte immunophenotyping. Effect of specimen age on results. *J Immunol Methods*, **165**, 31–35.

Owens AM, Vall GH, Hurley AA and Wormsley BS (2000). Validation and quality control of immunophenotyping in clinical flow cytometry. *J Immunol Methods*, **243**, 33–50.

Paciel J, Vignale RA, Bruno J, Vazquez G and De Anda G (1984). Increase of circulating HLA-DR+ T lymphocytes in psoriasis. *Med Cutan Iberto Lat Am*, **12**, 497–500.

Paxton H and Bendele T (1993). Effect of time, temperature, and anticoagulant on flow cytometry and hematological values. *Ann NY Acad Sci*, **677**, 440–443.

Porrini AM, Gambi D and Malatesta G (1992). Memory and naïve CD4+ lymphocytes in multiple sclerosis. *J Neurol*, **239**, 437–440.

Qin S, Rottman JB and Myers P (1998). The chemokine receptors CXCR3 and CCR5 mark subsets of T cells associated with certain inflammatory reactions. *J Clin Invest*, **101**, 746–754.

Rogge L, Barberis-Maino L and Biffi M (1997). Selective expression of an interleukin-12 receptor component by human T helper 1 cells. *J Exp Med*, **185**, 825–831.

Rose LM, Ginsburg AH, Rothstein TL, Ledbetter JA and Clark EA (1985). Selective loss of a subset of T helper cells in active multiple sclerosis. *Proc Natl Acad Sci USA*, **82**, 7389–7393.

Rottman JB, Smith TL, Ganley KG, Kikuchi T and Krueger JG (2001). Potential role of the chemokine receptors CXCR3, CCR4, and the integrin αEβ7 in the pathogenesis of psoriasis vulgaris, *Lab Invest*, **81**, 335–347.

Scolozzi R, Boccafogli A, Tola MR, Vicentini L, Camerani A, Degani D, Granieri E, Caniatti L and Paolino E (1992). T-cell phenotypic profiles in the cerebrospinal fluid and peripheral blood of multiple sclerosis patients. *J Neurol Sci*, **108**, 93–98.

Strauss K, Hulstaert F, Deneys V, Mazzon AM, Hannet I, De Bruyere M, Reichert T and Sindic CJ (1995). The immune profile of multiple sclerosis: T-lymphocyte effects predominate over all other factors in cyclophosphamide-treated patients. *J Neuroimmunol*, **63**, 133–142.

Takashima A and Morita A (1991). Genomic, phenotypic, and functional analyses of T cells in patients with psoriasis undergoing systemic cyclosporin A treatment. *J Invest Dermatol*, **96**, 376–382.

Tomkinson BE, Wagner DK, Nelson DL and Sullivan JL (1987). Activated lymphocytes during acute Epstein–Barr virus infection. *J Immunol*, **139**, 3802–3807.

Wadhwa M and Thorpe R (1998). Cytokine immunoassays: recommendations for standardisation, calibration and validation. *J Immunol Methods*, **219**:1–5.

7 T-cell surface markers in human peripheral whole blood using flow cytometry

Manjula Reddy[1], Cuc Davis, Hugh Davis, Charles Pendley and Uma Prabhakar
Centocor Research and Development, Inc., 145 King of Prussia Road, Radnor, PA 19087

7.1 Introduction

Evaluation of the expression of lymphocyte marker associated with autoimmune diseases is becoming increasingly important (Kagen et al., 2006, Mallone and Nepom, 2005, Mei et al., 2006, Sellebjerg et al., 2005, Sigmundsdottir et al., 2001). In our laboratory, wehave validated flow cytometric assays for several T-lymphocyte cell surface markers in whole blood samples such as IL-12 family receptors, Interleukin receptor β1 (IL-12Rβ1), Interleukin receptor β2 (IL-12Rβ2), and Interleukin

[1] To whom correspondence should be addressed.

Validation of Cell-Based Assays in the GLP Setting: A Practical Guide Editors, Uma Prabhakar and Marian Kelley

receptor 23 (IL-23R); activation markers, Interleukin-2 receptor α (CD25), CD69, and HLA-DR; chemokine receptors, CXCR3 and CCR5; co-stimulatory markers, CD28, CD80, and CD86; naïve (CD45RA) and memory (CD45RO) markers, and cutaneous lymphocyte antigen (CLA) on T-lymphocytes. The expression of these markers was evaluated in clinical samples from patients treated with immunomodulating agents. The validation of flow cytometric assays for use in clinical sample analysis is described in this chapter using IL-12Rβ1 and CLA markers as examples of markers that have high and low expression levels, respectively.

In patients with multiple sclerosis (MS), an autoimmune disease in which dysregulation of the IL-12 and IL-12 receptor (IL-12R) system on activated T lymphocytes has been reported (Ozenci et al., 2001), increased numbers of $CD25^+$ T cells in the peripheral blood of MS patients were observed (Porrini et al., 1992, Scolozzi et al., 1992). Further, changes in $CD4^+CD25^+$ cells was shown to correlatewith disease activity in patients with relapsing-remitting MS (Khoury et al., 2000, Porrini et al., 1992). In another Th1 type autoimmune disease, psoriasis, a skin homing molecule, CLA was reported to play a major role in disease pathogenesis (Berg et al., 1991, Rossiter et al., 1994, Sigmundsdottir et al., 2001). Increased CLA expression on peripheral blood T cells and psoriatic lesions was found to correlate with disease severity in psoriasis. Increased numbers of circulating $CD25^+$ T lymphocytes have also been seen in patients with active psoriasis (Ferenczi et al., 2000).

Flow cytometric assays have been used for investigational monitoring of the expression of IL-12R in peripheral whole blood samples, cerebrospinal fluid, synovial tissue, and intestinal lamina propria in patients with MS, rheumatoid arthritis, and Crohn's disease (Aita et al., 2004, Okazawa et al., 2002, Ozenci et al., 2001). Monitoring of CD25 expression in peripheral lymphocytes in patients with psoriasis and cutaneous lupus erythematosus (Sugiyama et al., 2005, Wenzel et al., 2005) and similarly, CLA expression in skin and peripheral blood of patients with psoriasis were also reported (Teraki et al., 2004). Validation of a flow cytometric assay in a GLP setting to measure changes in markers such as IL-12R, CD25, CLA, and CXCR3 in peripheral whole blood is described in this chapter and can be applied to monitor the effect of immune modulators in diseases such as MS or psoriasis.

Three- and four-color multi-fluorochrome antibody staining reagents are currently being used to immunotype lymphocytes in whole blood using flow cytometry (Keeney et al., 2004, Mandy et al., 2003, Owens et al., 2000). The acceptable reproducibility and precision of this technique in clinical laboratories come from careful validation and rigorous quality control (Keeney et al., 2004, Owens et al., 2000, Mandy et al., 2003). This chapter describes examples of the development, optimization and validation of flow cytometric assays that can be used to evaluate the expression of T-cell surface markers in peripheral whole blood in clinical study samples in a GLP setting.

7.2 Validation of cell surface marker expression in peripheral blood

Validation parameters included instrument performance, reagents, assay precision, and evaluating the stability of the marker expression to be tested. In addition, parameters important for activation markers, such as CD25 are also described. Because we intended to measure CLA, IL-12Rβ1, and CD25 expression in clinical study samples shipped from clinical sites, the stability of these markers was examined in more detail. Experiments were designed to simulate anticipated sample handling and testing in clinical studies. All subsequent validation experiments such as precision etc, were performed using 24 hours post-collection blood samples, to match the anticipated time it would take for patient blood samples drawn at clinical sites to reach a central clinical laboratory to be analyzed.

Instrument validation

Setup of the instrument and the daily quality control of light scatter and fluorescence measurements are major considerations for ensuring the proper functioning of the flow cytometer during validation and clinical sample testing. Because these procedures are instrument-specific, the manufacturer's instructions should be followed (Owens et al., 2000). The following validation was performed using a BD FACSCalibur™ flow cytometer (BD Biosciences, San Jose, CA). Therefore, the fluorescence and calibration check were performed using BD Calibrite beads (BD Biosciences, San Jose, CA). The BD instrument's FACSComp™ software automatically makes the necessary adjustments in instrument setup based on lot-specific performance ranges of the Calibrite beads. The FACSComp output of a successful calibration check is documented, and the records are stored for reviewing trends in laser power and voltage settings.

Reagent validation and specificity

The validation of cell surface marker expression should include appropriate combinations of various monoclonal antibodies labeled with different fluorochromes for multi-parameter analysis. It is important to select an appropriate clone of the antibodies as different clones may bind differently and show diverse staining intensity. For each multicolor antibody cocktail, the performance of each antibody when used alone or in combination should be validated. This process is discussed here using IL-12Rβ1 and CLA antibodies (Baumgarth and Roederer, 2000, Owens et al., 2000). The specificity of staining T lymphocytes with fluorochrome-labeled IL-12Rβ1 and CLA antibodies was validated by comparing the specificity of each antibody when it was used as a single reagent vs part of a cocktail with other multiple fluorochrome-conjugated antibodies.

The single-color IL-12Rβ1 antibody reagent was titrated to determine the test volume needed to achieve staining equivalent to that obtained with the multicolor cocktail. The antibody concentration recommended by the vendor was optimized to produce saturation staining of the IL-12Rβ1 marker. The percent of IL-12Rβ1$^+$ peripheral blood lymphocytes (PBLs) was determined from two healthy donors that were stained in duplicate with IL-12Rβ1 phycoerythrin (PE) antibody alone, IL-12Rβ1 PE in combination with CD3 peridinin chlorophyll protein (PerCP) as a dual-color reagent, or using four-color reagents. Three four-color reagents were used based on the anticipated population of interest to be analyzed: 1) IL-12Rβ1 PE in combination with CD28 fluorescein isothiocyanate (FITC), CD3 PerCP, and CD4 allophycocyanin (APC); 2) IL-12Rβ1 PE in combination with CD3 PerCP, CD4 APC, and CD8 FITC; or 3) IL-12Rβ1 PE in combination with CD3 PerCP, CD4 FITC, and CXCR3 APC (Table 7.1). Only the data for the third four-color antibody combination are shown in Table 7.1. A representative donor profile of IL-12Rβ1$^+$ is shown in Figure 7.1, Panel A. The percent of IL-12Rβ1$^+$cells stained by the fluorochrome-labeled antibody to IL-12Rβ1 was similar when single and multi-fluorochrome reagents were used (Table 7.1). The SD of the percent IL-12Rβ1$^+$ stained cells in both donors for single and multi-fluorochrome reagents was 1.83 to 4.92 and the %CV ranged from 3.50 to 17.12 (Table 7.1).

Similar experiments were conducted for the fluorochrome-labeled CLA antibody using peripheral blood lymphocytes (PBLsin whole blood from two healthy donors. To determine the percent of CLA$^+$ PBLs, samples were stained in duplicate with CLA FITC antibody alone or CLA FITC antibody in combination with CD3 PerCP as a dual-color reagent, in combination with CD3 PerCP and CD4 APC or CD8 PE as a three-color reagent (data not shown), or in combination with all three T-lymphocyte markers as a four-color reagent (Table 7.1). A representative donor profile of CLA$^+$is shown in Figure 7.1, Panel B. The percent of CLA$^+$ PBL staining with CLA FITC alone or in combination with other reagents was also similar (Table 7.1). The standard deviation (SD) of the percent CLA$^+$ stained cells for single and multi-fluorochrome reagents was $\leq$0.77 in both donors and the percent coefficient of variation (%CV) ranged from 1.27 to 6.25 (Table 7.1).

Table 7.1 Specificity of single and multiple-fluorochrome staining of peripheral blood CLA+ and IL-12Rß1+ lymphocytes*†

	IL-12Rß1+			CLA+		
	IL-12Rß1 PE	IL-12Rß1 PE CD3 PerCP	CD4 FITC IL-12Rß1 PE CD3 PerCP CXCR3 APC	CLA FITC	CLA FITC CD3 PerCP	CLA FITC CD8 PE CD3 PerCP CD4 APC
Donor 1	53.56	57.42	51.15	12.65	12.56	13.98
Donor 1	53.46	56.01	50.7	12.01	11.66	13.09
Mean (Donor 1)	53.51	56.72	50.93	12.33	12.11	13.54
Mean (Single and Multi)		55.11	52.22		12.22	12.33
SD (Single and Multi)		2.27	1.83		0.16	0.77
%CV (Single and Multi)		4.11	3.5		1.27	6.25
Donor 2	27.38	36.43	28.64	6.57	6.15	5.98
Donor 2	28.47	32.28	30.39	6.65	6.06	5.66
Mean (Donor 2)	27.93	34.36	29.52	6.61	6.11	5.82
Mean (Single and Multi)		31.14	28.72		6.36	6.03
SD (Single and Multi)		4.55	4.92		0.36	0.35
%CV (Single and Multi)		14.6	17.12		5.62	5.87

* Values are percent of positive lymphocytes.
†FITC indicates fluorescein isothiocyanate; PerCP, peridinin chlorophyll protein; PE, phycoerythrin; APC, allophycocyanin.

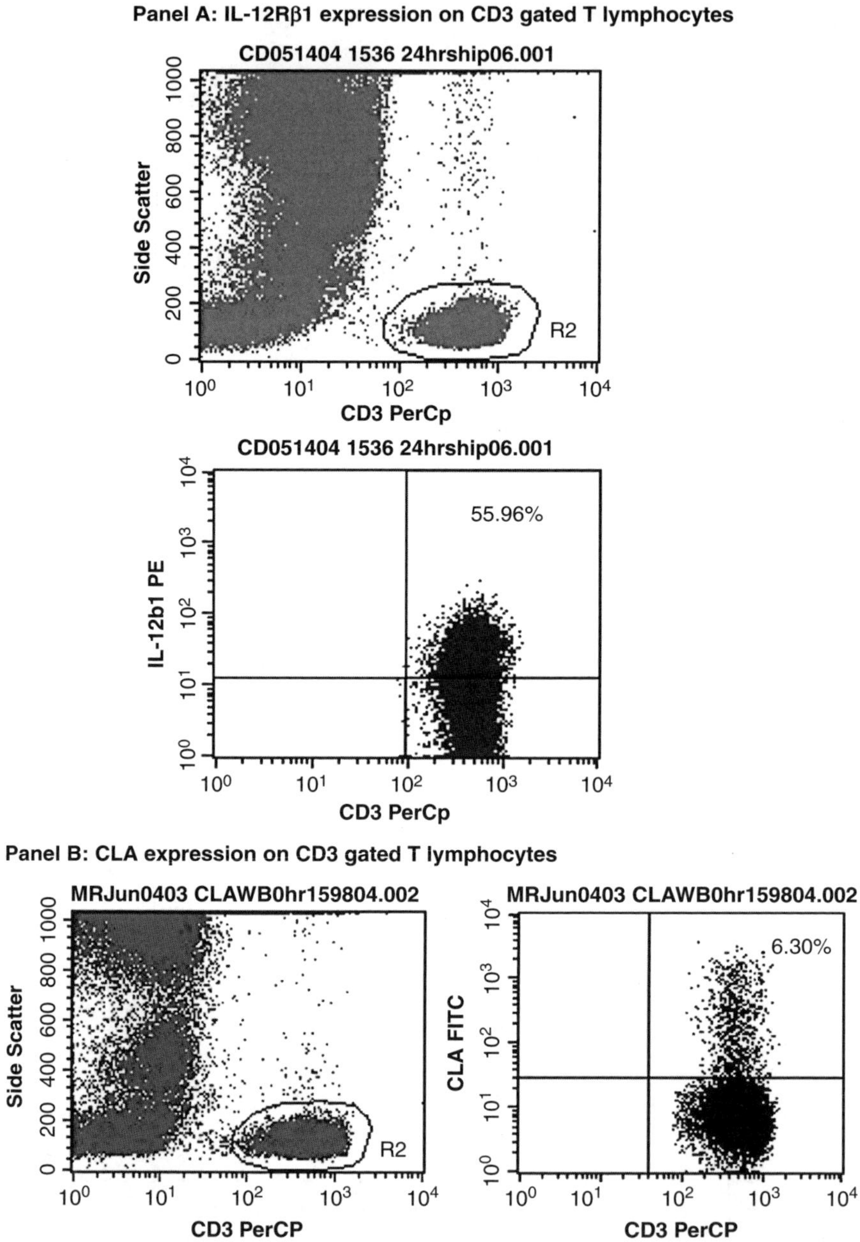

Figure 7.1 Immunophenotyping analysis of T lymphocytes. Panel A shows cells expressing IL-12Rβ1 and CD3 in a whole blood sample from a healthy human donor representing the profile of one of ten donors examined. The percentage of $CD3^{+}IL\text{-}12R\beta1^{+}$ T lymphocytes is visible in the upper right quadrant of the dot plot. Panel B shows cells expressing CLA and CD3 in a whole blood sample from a healthy human donor. The percentage of $CD3^{+}CLA^{+}$ T lymphocytes is visible in the upper right quadrant of the dot plot.

The presence of multiple antibodies did not interfere with the specificity of staining for the CLA and IL-12Rβ1 surface markers. Our findings indicate that percent of IL-12Rβ1$^+$ and CLA$^+$ T-lymphocyte subsets can be determined in whole blood using different multi-fluorochrome reagents depending on the T-lymphocyte population of interest. Based on these data, all subsequent experiments discussed in this chapter include determination of the percent of CLA$^+$ and IL-12Rβ1$^+$ T lymphocyte subsets using four-color reagents.

Intersubject range of marker expression

It is important to determine the baseline intersubject range of the expression of the T-cell surface markers both in normal volunteers and the subjects withthe disease state of interest in the assay validation. Cell surface marker expression in blood samples from treatment-naïve patients with psoriasis and MS, for example, should be compared with samples from healthy individuals.

Intersubject variability of the IL-12Rβ1 and CLA markers was evaluated by staining peripheral whole blood samples from ten healthy human donors with the IL-12Rβ1 or CLA antibody cocktails, respectively. CD3$^+$IL-12Rβ1$^+$ T cells ranged from 16.29% to 60.69%, and the percent of CD3$^+$CLA$^+$ cells ranged from 3.93% to 12.74% (confirming observation in an earlier report by Picker et al., 1990), indicating high variability of IL-12Rβ1 and CLA expression among healthy individuals. The intersubject variability analysis suggests that comparison within a patient but not between patients is valid.

Stability of marker expression during storage

The stability of the marker after blood draw until analysis is an essential aspect of validation. Blood samples from 5 to 10 healthy donors are tested for the expression of the marker immediately after blood draw and tested after storage or shipment based on the anticipated sample shipment conditions in a clinical study. The stored or shipped samples must be part of the same blood draw and be tested using the same reagents and protocol.

CLA stability in laboratory stored samples

The stability of CLA cell surface marker expression on human T lymphocytes was evaluated in whole blood collected from five normal donors after 0, 24, and 48 hours of storage at room temperature and at 4°C. The percentage of CLA$^+$ T lymphocytes remained stable in blood stored for 24 hours at room temperature in all five donors with an SD of $\leq$0.81 and a %CV from 3.36 to 13.90 (Table 7.2). At 4°C, the SD was $\leq$0.27 and the %CV 4.53 to 16.27 in four donors and 37.66 in one donor. At 48 hours, the SD of CLA$^+$ cells ranged from 0.47 to 1.76 in all five donors at both temperatures. The %CV ranged from 9.53 to 17.11 in four out of five donors (one donor showed %CV of 22.13) in 48-hour room temperature samples. In blood samples stored at 4°C for 48 hours, the %CV ranged from 2.26 to 27.74 (with one donor showing 44.02) (Table 7.2).

Table 7.2 Stability of $CD3^{+}CLA^{+}$ T lymphocytes in whole blood samples

Donor	0 Hours	24 Hours at RT	24 Hours at 4°C	48 Hours at RT	48 Hours at 4°C
Mean Donor 1	12.56	11.70	11.16	10.95	10.46
Mean (Before and After)		12.13	11.86	11.76	11.51
SD (Before and After)		0.61	0.99	1.14	1.49
%CV (Before and After)		5.04	8.38	9.68	12.93
Mean Donor 2	5.26	4.75	3.05	7.21	2.76
Mean (Before and After)		5.00	4.15	6.23	4.01
SD (Before and After)		0.36	1.56	1.38	1.76
%CV (Before and After)		7.14	37.66	22.13	44.02
Mean Donor 3	6.04	5.76	5.66	5.15	5.85
Mean (Before and After)		5.90	5.85	5.59	5.94
SD (Before and After)		0.20	0.27	0.63	0.13
%CV (Before and After)		3.36	4.53	11.26	2.26
Mean Donor 4	5.31	4.49	4.21	4.64	3.57
Mean (Before and After)		4.90	4.76	4.97	4.44
SD (Before and After)		0.58	0.77	0.47	1.23
%CV (Before and After)		11.77	16.27	9.53	27.74
Mean Donor 5	6.37	5.23	5.27	5.00	4.71
Mean (Before and After)		5.80	5.82	5.68	5.54
SD (Before and After)		0.81	0.78	0.97	1.18
%CV (Before and After)		13.90	13.43	17.11	21.26

* Values are percent of positive lymphocytes.
† ON indicates overnight; RT, room temperature.

CLA^+ T lymphocyte subsets can be determined with greater accuracy when analyzed the day after blood collection. Because there is greater variability at 48 hours, samples stored or received on the second day after collection will be excluded from analysis for this marker. Furthermore, storing blood samples at 4°C until analysis did not enhance the stability of CLA and is therefore not required.

CD25 stability in laboratory stored samples

The temperature stability of CD25 cell surface marker expression on human T lymphocytes was evaluated in whole blood samples collected from five healthy donors following 0 and 24 hours of laboratory storage at room temperature. The percentage of T lymphocytes expressing CD25 antigen in all five donors remained stable after storage for 24 hours at room temperature compared to before storage with an SD $\leq$2.38 and a %CV range of 1.14 to 11.02 (Table 7.3C).

Stability of marker expression during shipment

IL-12Rβ1 stability in shipped samples

The stability of the expression of the IL-12Rβ1 cell surface marker in human T lymphocytes was evaluated in whole blood samples before shipment, after overnight storage in the laboratory at room temperature, or following overnight or 2-day shipment at room temperature. Samples were collected from five healthy donors. The percent of T lymphocytes expressing IL-12Rβ1 in all five donors remained stable in blood samples stored in the laboratory overnight at room temperature, with SDs ranging from 0.10 to 6.98 and with %CVs ranging from 0.32 to 16.82 (Table 7.3A). Blood samples from all donors shipped overnight at room temperature had a range of SD from 0.19 to 5.66 and a range of %CV from 0.77 to 13.33. Samples received on the second-day post blood draw had an SD range of 0.49 to 5.21 and %CV range of 1.65 to 12.20. Based on these results, the percent of IL-12Rβ1$^+$ T-lymphocyte subsets can be accurately determined in blood samples received 2 days after collection and shipment at room temperature.

CLA stability in shipped samples

The effect of shipment temperature on the stability of CLA cell surface marker expression in overnight shipment at room temperature or on ice packs, as well as after storage overnight in the laboratory at room temperature was evaluated. Samples collected from five healthy donors were used

Table 7.3A Stability of IL-12Rβ1 cell surface marker expression on $CD3^{+}IL\text{-}12R\beta1^{+}$ T lymphocytes in whole blood samples shipped overnight*†

Donor	Before Shipment	After ON RT Storage	After ON Shipment at RT	After 2-Day Shipment at RT
Mean Donor 1	60.69	56.82	55.44	58.95
Mean (Before and After)		58.75	58.07	59.82
SD (Before and After)		2.74	3.71	1.23
%CV (Before and After)		4.66	6.39	2.06
Mean Donor 2	46.45	36.57	38.45	39.07
Mean (Before and After)		41.51	42.45	42.76
SD (Before and After)		6.98	5.66	5.21
%CV (Before and After)		16.82	13.33	12.20
Mean Donor 3	29.51	30.24	30.02	35.08
Mean (Before and After)		29.88	29.76	32.30
SD (Before and After)		0.52	0.36	3.94
%CV (Before and After)		1.73	1.20	12.20
Mean Donor 4	25.00	23.60	24.73	22.72
Mean (Before and After)		24.30	24.87	23.86
SD (Before and After)		0.99	0.19	1.61
%CV (Before and After)		4.07	0.77	6.76
Mean Donor 5	30.05	29.91	31.47	29.35
Mean (Before and After)		29.98	30.76	29.70
SD (Before and After)		0.10	1.00	0.49
%CV (Before and After)		0.32	3.26	1.65

* Values are percent of positive lymphocytes.
†ON indicates overnight; RT, room temperature.

for these experiments. The results are shown in Table 7.3B. The percent of T lymphocytes expressing CLA antigen remained stable in shipped samples at room temperature or with an ice pack. An SD $\leq$1.19 was calculated for all five donors for both conditions, with a %CV range of 0.00 to 13.47 in samples from four of five donors shipped overnight at ambient temperature (23.08 in the other donor), and 0.12 to 7.73 in all donor samples shipped overnight on ice pack. These results indicate that shipment conditions do not affect the stability of the expression of the CLA marker and that its surface expression can be determined accurately in whole blood samples shipped overnight from clinical sites at both ambient temperature and on ice packs. Shipment at cold temperature appeared to decrease variability in three of the donors tested in this experiment. However, since the analysis of CLA marker expression may be performed along with the analysis of other markers that might be sensitive to cold temperature, clinical blood

Table 7.3B Stability of CLA cell surface marker expression on $CD3^+CLA^+$ T lymphocytes in whole blood samples shipped overnight*†

Donor	Before Shipment	After ON RT Storage	After ON Shipment at RT	After ON Shipment on Ice Pack
Mean Donor 1	11.28	11.53	12.56	12.59
Mean (Before and After)		11.41	11.92	11.93
SD (Before and After)		0.18	0.91	0.92
%CV (Before and After)		1.55	7.59	7.73
Mean Donor 2	4.32	5.61	6.01	4.51
Mean (Before and After)		4.96	5.16	4.41
SD (Before and After)		0.91	1.19	0.13
%CV (Before and After)		18.31	23.08	2.96
Mean Donor 3	5.49	5.73	6.12	5.46
Mean (Before and After)		5.61	5.80	5.47
SD (Before and After)		0.17	0.45	0.02
%CV (Before and After)		3.03	7.74	0.32
Mean Donor 4	3.04	3.42	3.68	3.05
Mean (Before and After)		3.23	3.36	3.04
SD (Before and After)		0.27	0.45	0.00
%CV (Before and After)		8.32	13.47	0.12
Mean Donor 5	4.00	3.88	4.00	3.97
Mean (Before and After)		3.94	4.00	3.99
SD (Before and After)		0.08	0.00	0.02
%CV (Before and After)		2.15	0.00	0.53

* Values are percent of positive lymphocytes.
† ON indicates overnight; RT, room temperature.

samples will have to be shipped at ambient temperature when a combined analysis of all markers is planned.

CD25 stability in shipped samples

Stability over time and at different storage temperatures

The stability of activation marker CD25 on $CD3^+$ human T lymphocytes was evaluated over time and at different shipping temperatures. It is believed that makers such as those that represent cellular activation might be more sensitive to the temperature during blood storage and shipment.

Stability in "on-site" collected and overnight shipped whole blood

The effect of shipment on the stability of CD25 cell surface marker expression on $CD3^+$ human T lymphocytes was evaluated by staining whole blood collected from five healthy donors before shipment and after shipment in different conditions (Table 7.3C). Samples shipped and tested on the same day of collection had an SD range of 0.07 to 2.19 and a %CV range of 0.28 and 6.44 in all of the healthy donor samples (Table 7.3C). In samples shipped overnight at ambient temperature, the SD ranged from 0.03 to 4.57 and the %CV ranged from 0.07 to 3.48 in three of five donors, whereas the %CVs were 15.15 and 18.83 in two of five donors. Samples shipped overnight on ice packs had a SD range from 0.70 to 1.76, and a %CV range from 0.63 to 5.04 in all five donors (Table 7.3C). These results were relatively less variable compared with the results for shipment at room temperature.

The results of these experiments indicated that either same day or overnight shipment had minimal impact on the detection of the percent of $CD3^+CD25^+$ T lymphocytes in whole blood samples. The CD25 antigen density was also examined on $CD3^+$ T lymphocytes in shipped samples and results were compared with those obtained before shipment (data not shown). Because variability of antigen density increased with time, samples received on the second day of shipment should be excluded from analysis. Furthermore, shipment on ice packs did not greatly enhance CD25 marker stability and is not recommended. The percent of $CD25^+$ T lymphocytes and the density of the antigen can be analyzed accurately in whole blood samples shipped at ambient temperatures from clinical sites.

It was evident from these studies that the population of $CD3^+CD25^+$ T-lymphocyte subsets can be determined with greater accuracy in whole blood when samples are analyzed within 24 hours after collection. Clinical samples received on the second day after collection should be excluded from analysis for this marker because of greater variability in the results at 48 hours after sample collection. Furthermore, storage of the samples at 4°C until analysis did not enhance the stability of CD25 and is therefore not required.

Stability of marker expression over time

Another important validation parameter, the physiological variability of the marker over time should be assessed. The time period of the stability studies should encompass the time length of the clinical study in which the marker will be assessed. The long-term stability of the expression of the IL-12Rβ1 T-cell surface marker was examined in whole blood from five donors over a 16-week period (Table 7.4). In healthy volunteers, the levels of IL-12Rβ1 expression remained relatively stable over the time course monitored with SD ranging from 1.29 to 5.47 and %CV ranging from 6.96 to 26.16. Variability in expression of the CLA marker on T cells in peripheral whole blood was also analyzed over a 7-week period in three healthy donors who had high, moderate, and low levels of CLA expression

Table 7.3C Stability of CD25 cell surface marker expression on $CD3^{+}CD25^{+}$ T lymphocytes in shipped whole blood samples*†

Donor	Before Shipment	After Same Day Shipment	After ON Laboratory Storage	After ON Shipment at RT	After ON Shipment on Ice Pack
Mean Donor 1	43.38	44.26	42.69	43.42	45.87
Mean (Before and After)		43.82	43.03	43.40	44.62
SD (Before and After)		0.62	0.49	0.03	1.76
%CV (Before and After)		1.42	1.14	0.07	3.94
Mean Donor 2	27.53	29.01	28.36	21.06	29.57
Mean (Before and After)		28.27	27.94	24.30	28.55
SD (Before and After)		1.05	0.58	4.57	1.44
%CV (Before and After)		3.70	2.09	18.83	5.04
Mean Donor 3	25.43	25.33	24.47	20.51	25.21
Mean (Before and After)		25.38	24.95	22.97	25.32
SD (Before and After)		0.07	0.68	3.48	0.16
%CV (Before and After)		0.28	2.72	15.15	0.63
Mean Donor 4	18.52	19.10	21.65	18.00	17.54
Mean (Before and After)		18.81	20.09	18.26	18.03
SD (Before and After)		0.41	2.21	0.37	0.70
%CV (Before and After)		2.16	11.02	2.01	3.86
Mean Donor 5	32.42	35.52	35.79	33.42	34.28
Mean (Before and After)		33.97	34.10	32.92	33.35
SD (Before and After)		2.19	2.38	0.70	1.31
%CV (Before and After)		6.44	6.98	2.14	3.93

* Values are percent of positive lymphocytes.

† ON indicates overnight; RT, room temperature.

(Table 7.4). The levels of CLA expression remained relatively stable over 7 weeks in each of these healthy donors with an SD of ≤0.86 and a range of %CV from 5.33 to 13.82. Because of the variable levels of expression of IL-12Rβ1 and CLA in whole blood in different donors, the levels of IL-12Rβ1 and CLA expression in each patient at various time points should be compared with individual baseline values in clinical studies (pre-treatment vs post-treatment comparison).

Precision of the assay

Inter-assay precision

Whole blood samples from two healthy donors were stained in triplicate in three separate experiments to evaluate inter-assay variability of CLA and IL-12Rβ1 marker detection. The measurement of the percent of positively stained $CD3^+CLA^+$ and $CD3^+$IL-12Rβ1$^+$ T lymphocytes was highly precise between assays as indicated by a pooled SD of ≤0.21 and a %CV range from 2.64 to 4.32 for CLA and a pooled SD of ≤0.43 and %CV of ≤1.22 for IL-12Rβ1 in both donors (Table 7.5).

Intra-assay precision

Whole blood samples from two healthy donors were stained in triplicate to evaluate variability within the assay. The SD of $CD3^+$ T lymphocytes staining positively for the CLA T-lymphocyte surface marker was ≤0.22 with a %CV ≤3.91 for both donors (Table 7.5). The SD of percentage of positive $CD3^+$IL-12Rβ1$^+$ T lymphocytes was ≤0.51 with a %CV ≤1.44 (Table 7.6). These results indicate that intra-assay variability was acceptably low. The high assay precision justifies the analysis of clinical samples in duplicate rather than in triplicate.

Table 7.4 Variability of CLA and IL-12Rß1 marker expression on T lymphocytes over time*†

Donor	IL-12Rβ1 Expression Over 16 Weeks			CLA Expression Over 7 Weeks		
	Mean	SD	%CV	Mean	SD	%CV
1	34.08	2.38	6.98	12.22	0.65	5.33
2	14.48	1.29	8.94	6.23	0.86	13.82
3	44.33	3.08	6.96	4.50	0.52	11.60
4	20.89	5.47	26.16	ND	ND	ND
5	32.21	2.76	8.57	ND	ND	ND

* Values are percent of positive lymphocytes.
†ND indicates not determined.

Table 7.5 Inter-assay variability of positive staining of IL-12Rβ1 and CLA on T-lymphocyte subsets*

	IL-12Rβ1$^+$		CD3$^+$CLA$^+$	
	Donor 1	Donor 2	Donor 1	Donor 2
Assay 1	55.24	35.19	4.54	8.03
Assay 2	54.82	35.12	4.17	7.76
Assay 3	54.41	35.90	4.34	7.62
Mean of assays	54.82	35.40	4.35	7.80
SD	0.42	0.43	0.19	0.21
%CV	0.76	1.22	4.32	2.64

* Values are percent of positive lymphocytes.

Table 7.6 Intra-assay variability of positive staining of IL-12Rβ1 and CLA on T lymphocyte subsets*

	Replicates	CD3$^+$IL-12Rβ1$^+$	CD3$^+$CLA$^+$
Donor 1	1	54.11	4.51
Donor 1	2	54.70	4.57
Donor 1	3	54.53	4.85
Mean Donor 1		54.45	4.64
SD		0.30	0.18
% CV		0.56	3.91
Donor 2	1	36.38	7.85
Donor 2	2	35.42	8.20
Donor 2	3	35.58	7.80
Mean Donor 2		35.79	7.95
SD		0.51	0.22
% CV		1.44	2.74

* Values are percent of positive lymphocytes.

Inter-operator precision

Inter-operator variability of the procedure for staining T lymphocytes in whole blood with the fluorochrome-labeled antibody reagents was evaluated between two trained individuals who performed this procedure in duplicate on the same day using samples from the same two human donors. The results for each operator are shown in Table 7.7. The grand SD of the percent of CD3$^+$CLA$^+$ T lymphocytes obtained by the two operators was $\leq$0.29 in both donors and the grand %CV was 0.85 in one donor and 3.89 in the second donor. The grand SD and grand %CV of the percent CD3$^+$IL-12Rβ1$^+$ T lymphocytes obtained by the two operators in both donors (pooled) were $\leq$1.14 and $\leq$2.82, respectively. These results confirm that the procedure for staining T lymphocytes in whole blood with the

Table 7.7 Inter-operator variability of percent positive T lymphocytes expressing the IL-12Rβ1 and CLA cell surface markers*

	$CD3^+$IL-12Rβ1^+		$CD3^+CLA^+$	
	Donor 1	Donor 2	Donor 1	Donor 2
Operator 1	37.04	24.91	4.51	7.74
Operator 1	39.85	24.55	4.57	7.50
Mean of assays	38.45	24.73	4.54	7.62
SD	1.99	0.25	0.04	0.17
%CV	5.17	1.03	0.93	2.23
Operator 2	42.69	21.69	4.81	7.98
Operator 2	42.16	21.14	4.95	7.15
Mean of assays	42.43	21.42	4.88	7.57
SD	0.37	0.39	0.10	0.59
%CV	0.88	1.82	2.03	7.76
Grand Mean	40.44	23.07	4.71	7.59
Grand SD	1.14	0.10	0.04	0.29
Grand %CV	2.82	0.41	0.85	3.89

* Values are percent of positive lymphocytes.

fluorochrome-labeled reagents has high precision when trained operators perform the procedure.

7.3 Sample and Data Acceptance Criteria

Sample and data acceptance criteria should follow the assay validation. Listed below are the established criteria for the analysis of the expression of IL-12Rβ1, CLA, and CD25 in T lymphocytes in clinical study samples based on the results of the validation studies described in this chapter.

- The expression of IL-12Rβ1 can be analyzed in clinical blood samples shipped at room temperature that are received on the next or second day after collection. Other markers in the samples may not be stable up to 2 days after collection. Because expression of these less stable markers may be assessed along with IL-12Rβ1 expression, the analysis of IL-12Rβ1 expression will be performed only in samples received the next day after collection. Any samples received outside of this time frame will not be processed.
- The expression of CLA and CD25 will be analyzed in clinical samples received the day after collection. Any samples received 2 days after collection will be excluded from the analysis.

- Samples shipped on ice packs that are received the day after collection can be used for CLA and CD25 marker analysis. Cold-sensitive markers may not be able to be analyzed in these samples.
- Samples will be tested in duplicate based on the high precision of the assay. If high variability between duplicates occurs due to operator error or instrument error (incorrect staining, aspiration of sample, low event count, etc.), the duplicates will not be averaged and the appropriate replicate, based on the sample scatter properties and staining profile, will be reported.
- Results will be expressed mainly as percent positive IL-12Rβ1, CLA, and CD25 lymphocyte subsets in samples before and during and/or after treatment. Data from clinical samples will further be expressed as percent change from baseline for each patient.
- Results can also include level of IL-12Rβ1, CLA, and CD25 expression (mean fluorescence intensity [MFI] values of pre- and post-treatment) if necessary, based on the treatment and study protocol.

7.4 Implementation of the Validated Assay

An approved validation report that includes all the information mentioned above and an assay checklist that includes information on reagents and equipment along with various steps in the assay method is recommended prior to initiation of clinical sample analysis.

The following procedures are used to implement the assay for a clinical study

- The optical alignment, fluorescence resolution, and fluorescence intensity are checked on a daily basis to confirm instrument performance before any clinical samples are analyzed and documented in the logbook
- Instrument setup and optimization are performed utilizing procedures specific for the instrument such as Calibrite beads and FACSComp software or BD FACSCalibur.
- The lymphocyte population is identified and gated (default label R1). Determine the level of background staining based on the isotype control by adjusting voltages of the FL1, FL2, FL3, and FL4 detectors.
- The compensation for the T-cell surface markers being analyzed is adjusted.
- The instrument settings are saved after the appropriate adjustments are made.
- Sample data are acquired by collecting 20,000 events in the lymphocyte gate.
- Controls samples such as CD-Chex Plus (Streck Inc., 2006) used for instrument and reagent performance are stained with the same reagents

used for test samples and analyzed using the same methods. Because the CD-Chex PLUS samples may not contain all the surface markers being tested in the test samples, the CD-Chex PLUS should be stained with the positive control antibody panel that is part of the test sample panel, for example, a T cell panel CD3/CD45/CD4/CD8 and /or a T, B, natural killer cell panel, CD3/CD45/CD16+CD56/CD19. The vendor recommends each lab establish their own lot specific range because staining procedures and reagents used by different laboratories could be different than those used by the vendor. Varying gating strategies, based on cell populations of interest, used at different laboratories also warrant establishment of laboratory-specific ranges. The percent positive cells from runs on different days are checked against the expected range for that lot of control blood sample and a run is successful if the control cell lymphocyte percentages fall within the vendor expected range.

7.5 Summary

The validation of three- and four-color flow cytometric assays for determining the expression of cell surface markers on T lymphocytes in human peripheral whole blood using GLP methods has been described using IL-12Rβ1, CLA, and CD25 as examples. The four-color fluorochrome antibody cocktail used to determine the percent of IL-12β1$^+$ lymphocyte subsets consisted of CD28 FITC, IL-12Rβ1 PE, CD3 PerCP, and CD4 APC. Although not described in this chapter, this reagent can also be used to validate an assay of the expression of CD28 on T lymphocytes in human peripheral whole blood samples. A different four-color fluorochrome antibody cocktail consisting of CLA FITC, CD8 PE, CD3 PerCP, and CD4 APC was used in the assay for the expression of CLA. A flow cytometric method was validated for CD25$^+$ T cells that used a three-color reagent consisting of CD4 FITC, CD25 PE, and CD3 PerCP. The accuracy and specificity of the staining using the three-color reagent was confirmed by comparison with results for the single-color antibody reagents specific for IL-12Rβ1, CLA, or CD25. The inter-assay variability of the IL-12Rβ1 and CLA staining was low suggesting a high level of assay precision that permitted analyzing patient samples in duplicate. The intra-assay variability was determined to be low as well. Minimal inter-operator variability confirmed the precision of the staining procedure for IL-12Rβ1 and CLA when performed by trained operators. The final results from clinical study samples will be expressed as percent change from baseline.

Clinical study samples may not be analyzed immediately after collection because they might have to be shipped from clinical study sites to the analytical laboratory. An important part of the validation consisted of stability assessment of the expression of the T-lymphocyte cell surface markers with time after sample collection and at various shipping temperatures. The CLA surface marker was stable up to 24 hours after collection in whole blood samples shipped overnight or stored up to 24 hours in the laboratory after collection. It was concluded that the percent of CLA^+ T lymphocytes could be assayed in whole blood samples from clinical studies collected at clinical study sites and shipped to the analytical laboratory as long as samples arrived the same day or the day after collection and were analyzed immediately upon receipt. The expression of the IL-12Rβ1 surface marker was stable in blood samples received the next day or the second day after collection and in blood stored overnight in the laboratory. CD25 surface marker expression was stable in whole blood samples after overnight shipment or stored in the laboratory for up to 24 hours.

In conclusion, this chapter provided an overview of parameters that need to be addressed during the validation of a flow cytometric assay for cell surface markers on peripheral blood lymphocytes in clinical samples. The validation parameters included: 1) instrument performance; 2) reagents, including reagent specificity; 3) the intersubject range and variability of marker expression; 4) the stability of the expression of the marker of interest over time; 5) the stability of the markers during shipment of the samples from the clinical sites to the analytical laboratory; and 6) the precision of the assay. Criteria for the acceptance of samples and data were defined, based on the results of the validation experiments. In keeping with working in a GLP setting, all validation data must be summarized in a validation report. Standard operating procedures for instrument identification and operation and assay checklists that include reagent lot numbers, the operators, sample identification, and documentation of completed steps for the analytical assay must be generated. The procedures and experiments described in this chapter for cell surface markers on T lymphocytes in peripheral whole blood can be applied to the validation of flow cytometric assays for other cell types.

References

Aita, T., Yamamura, M., Kawashima, M., Okamoto, A., Iwahashi, M., Yamana, J. & Makino, H. 2004. Expression of interleukin 12 receptor (IL-12R) and IL-18R on CD4+ T cells from patients with rheumatoid arthritis. *J Rheumatol*, vol 31, no 3, pp 448–56.

Berg, E. L., Yoshino, T., Rott, L. S., Robinson, M. K., Warnock, R. A., Kishimoto, T. K., Picker, L. J. & Butcher, E. C. 1991. The cutaneous lymphocyte antigen is a skin lymphocyte homing receptor for the vascular lectin endothelial cell-leukocyte adhesion molecule 1. *J Exp Med*, vol 174, pp 1461–6.

Ferenczi, K., Burack, L., Pope, M., Krueger, J. G. & Austin, L. M. 2000. CD69, HLA-DR and the IL-2R identify persistently activated T cells in psoriasis vulgaris lesional skin: blood and skin comparisons by flow cytometry. *J Autoimmun*, vol 14, no 1, pp 63–78.

Kagen, M. H., Mccormick, T. S. & Cooper, K. D. 2006. Regulatory T cells in psoriasis. *Ernst Schering Res Found Workshop*, no 56, pp 193–209.

Keeney, M., Barnett, D. & Gratama, J. W. 2004. Impact of standardization on clinical cell analysis by flow cytometry. *J Biol Regul Homeost Agents*, vol 18, pp 305–12.

Khoury, S. J., Guttmann, C. R., Orav, E. J., Kikinis, R., Jolesz, F. A. & Weiner, H. L. 2000. Changes in activated T cells in the blood correlate with disease activity in multiple sclerosis. *Arch Neurol*, vol 57, no 8, pp 1183–9.

Mallone, R. & Nepom, G. T. 2005. Targeting T lymphocytes for immune monitoring and intervention in autoimmune diabetes. *Am J Ther*, vol 12, no 6, pp 534–50.

Mandy, F. F., Nicholson, J. K. & Mcdougal, J. S. 2003. Guidelines for performing single-platform absolute CD4+ T-cell determinations with CD45 gating for persons infected with human immunodeficiency virus. Centers for Disease Control and Prevention. *MMWR Recomm Rep*, vol 52, pp 1–13.

Mei, F. J., Osoegawa, M., Ochi, H., Minohara, M., Nan, S., Murai, H., Ishizu, T., Taniwaki, T. & Kira, J. 2006. Long-term favorable response to interferon beta-1b is linked to cytokine deviation toward the Th2 and Tc2 sides in Japanese patients with multiple sclerosis. *J Neurol Sci*, vol 246, no 1-2, pp 71–7.

Okazawa, A., Kanai, T., Watanabe, M., Yamazaki, M., Inoue, N., Ikeda, M., Kurimoto, M., Ishii, H. & Hibi, T. 2002. Th1-mediated intestinal inflammation in Crohn's disease may be induced by activation of lamina propria lymphocytes through synergistic stimulation of interleukin-12 and interleukin-18 without T cell receptor engagement. *Am J Gastroenterol*, vol 97, no 12, pp 3108–17.

Owens, M. A., Vall, H. G., Hurley, A. A. & Wormsley, S. B. 2000. Validation and quality control of immunophenotyping in clinical flow cytometry. *J Immunol Methods*, vol 243, pp 33–50.

Ozenci, V., Pashenkov, M., Kouwenhoven, M., Rinaldi, L., Soderstrom, M. & Link, H. 2001. IL-12/IL-12R system in multiple sclerosis. *J Neuroimmunol*, vol 114, pp 242–52.

Picker, L. J., Terstappen, L. W., Rott, L. S., Streeter, P. R., Stein, H. & Butcher, E. C. 1990. Differential expression of homing-associated adhesion molecules by T cell subsets in man. *J Immunol*, vol 145, pp 3247–55.

Porrini, A. M., Gambi, D. & Malatesta, G. 1992. Memory and naive CD4+ lymphocytes in multiple sclerosis. *J Neurol*, vol 239, no 8, pp 437–40.

Rossiter, H., Van Reijsen, F., Mudde, G. C., Kalthoff, F., Bruijnzeel-Koomen, C. A., Picker, L. J. & Kupper, T. S. 1994. Skin disease-related T cells bind to endothelial selectins: expression of cutaneous lymphocyte antigen (CLA) predicts E-selectin but not P-selectin binding. *Eur J Immunol*, vol 24, pp 205–10.

Scolozzi, R., Boccafogli, A., Tola, M. R., Vicentini, L., Camerani, A., Degani, D., Granieri, E., Caniatti, L. & Paolino, E. 1992. T-cell phenotypic profiles in the cerebrospinal fluid and peripheral blood of multiple sclerosis patients. *J Neurol Sci*, vol 108, no 1, pp 93–8.

Sellebjerg, F., Ross, C., Koch-Henriksen, N., Sorensen, P. S., Frederiksen, J. L., Bendtzen, K. & Sorensen, T. L. 2005. CD26 + CD4 + T cell counts and attack risk in interferon-treated multiple sclerosis. *Mult Scler*, vol 11, no 6, pp 641–5.

Sigmundsdottir, H., Gudjonsson, J. E., Jonsdottir, I., Ludviksson, B. R. & Valdimarsson, H. 2001. The frequency of CLA+ CD8+ T cells in the blood of psoriasis patients correlates closely with the severity of their disease. *Clin Exp Immunol*, vol 126, pp 365–9.

Streck Inc. 2006. CD-Chex Plus. Available at: http://www.streck.com/products/immunology/cdchex_plus.asp. Accessed September 13, 2006.

Sugiyama, H., Gyulai, R., Toichi, E., Garaczi, E., Shimada, S., Stevens, S. R., Mccormick, T. S. & Cooper, K. D. 2005. Dysfunctional blood and target tissue CD4+CD25high regulatory T cells in psoriasis: mechanism underlying unrestrained pathogenic effector T cell proliferation. *J Immunol*, vol 174, no 1, pp 164–73.

Teraki, Y., Miyake, A., Takebayashi, R. & Shiohara, T. 2004. Homing receptor and chemokine receptor on intraepidermal T cells in psoriasis vulgaris. *Clin Exp Dermatol*, vol 29, pp 658–63.

Wenzel, J., Henze, S., Brahler, S., Bieber, T. & Tuting, T. 2005. The expression of human leukocyte antigen-DR and CD25 on circulating T cells in cutaneous lupus erythematosus and correlation with disease activity. *Exp Dermatol*, vol 14, no 6, pp 454–9.

8 Intracellular cytokine detection by flow cytometry

Julie G. Wilkinson[1], Carlos A. Aparicio[1] and Wade E. Bolton[2]

[1]*Custom BioPharma Solutions, Beckman Coulter, Inc., 11800 SW 147th Avenue, Miami, FL 33196, USA*

[2]*Custom Bio/Pharma Solutions, Beckman Coulter, Inc., 4300 N. Harbor Blvd., (M/C E-34-E), Fullerton, CA 92835, USA*

8.1 Introduction

Surrogate endpoints in oncology drug development include tumor response and quality of life, which are hardly ideal biomarkers. Better biomarkers would include changes in: genome: e.g., mutations, expression, function; proteome: e.g., expression, function, post-translational modifications (PTM); and the cytome: e.g., tissue histology, cellular function (proliferation, activation, apoptosis, differentiation). Cytokines are major players in health and disease, and the ability to quantitate them reproducibly may allow their use as biomarkers for specific diseases and therapies. When designed and applied under good laboratory practice (GLP) guidelines, intracellular cytokine staining (ICS) may demonstrate utility for defining

Validation of Cell-Based Assays in the GLP Setting: A Practical Guide Editors, Uma Prabhakar and Marian Kelley

mechanisms of action and surrogate markers of clinical outcome, and in monitoring drug efficacy.

Two multi-site studies for standardizing functional cellular assays elegantly highlight the immense effort required to validate these types of assays. The Enzyme-linked Immunospot (ELISPOT) proficiency panel conducted a study in 11 experienced laboratories participating in international HIV-1 vaccine trials with the goal of assessing laboratory competency and comparability when performing these assays for estimating the number of cells secreting IFNγ in response to specific antigens (Cox et al, 2005). There was a centralized peripheral blood mononuclear cells (PBMC) freeze and distribution while individual lab reagents and protocols were used with just a few group guidelines. There was good concordance of qualitative data, but this was not true for the quantitative data. The conclusion was that there is a need for better standardization of protocols and reagents to obtain reliable and reproducible data.

The NIAID–CANVAC interlab precision study for the standardization of ICS assays (Maecker et al, 2005) was another effort to characterize an assay being used to compare the immunogenicity of vaccine candidates across multiple international organizations conducting clinical trials. The study involved 15 experienced labs involved in HIV clinical trials. Testing was performed at all sites using common experimental protocols and common reagents. The conclusions were that a common protocol must be in place and followed strictly, that protocol reagents should be prepared centrally when possible, and that centralized analysis, or use of a standardized gating scheme of the cytometry data reduces variation. Assay precision was dependent on the percent of positive events: with >0.5% IFNγ+ T cells the coefficient of variation (CV) = 18–24%, and with <0.1% IFNγ+ T cells the CV increased to 57–82%.

The focus of this chapter will be on the design, optimization, verification, and validation of a standardized in vitro test system for the reproducible measurement of intracellular cytokines (ICS) through control of variables and automation.

8.2 Assay standardization

The majority of bioassays used in research labs, regardless of how faithfully they are performed, will suffer failures when used to support a clinical trial. This is largely due to the increased scale of specimen testing which necessitates multiple lots of every reagent, multiple enrolment sites, multiple testing sites, multiple technicians, multiple instruments, multiple analysts, and multiple time points extended over several years

The initial establishment of GLP guidelines for day-to-day laboratory work is time consuming, and each laboratory must apply the principles in a way that complements their facility structure and infrastructure. There are commonalities, however, across laboratories performing in vitro assays, specifically the need to define the assay test system itself and to design quality assurance and control into the system. Table 8.1 summarizes the GLP components and how each one applies to standardization of a functional cell-based assay such as intracellular cytokine staining.

Table 8.1 GLP components of functional cell-based assays

	GLP Component	Application to an *in vitro* ICS assay
Quality Assurance	Approved Standard Operating Procedures (SOP)	1. Specimen Collection 2. Specimen Shipping, Storage, and Processing 3. Sample Preparation 4. Data tracking, acquisition, analysis & management 5. Instrument Validation 6. Lab Operations (e.g. equipment certification/calibration, liquid nitrogen facility monitoring, etc.) 7. Reagent/Material Certifications 8. Personnel Training Requirements and Certification
	Statistical Procedures	1. Collaboration with a Biostatistician is critical. 2. Design specific statistical analysis procedures for defining detection limits, confidence intervals, etc. 3. Regulatory agencies may describe acceptable statistical procedures. 4. For designing validation testing, there are guideline "magic" numbers. 3 and 10 are for the number of sites and number of replicates per condition. 50 and 100 are the total number of donors/patients required
	Instrument Validation	1. Establish objective specifications that define acceptable operation of all analytical laboratory instruments 1a. Define "control limits" assigned as upper and lower ranges around the expected instrumental output 1b. Consult instrument manufacture's or manuals 2. Continuously assess the validity of instrument data

Table 8.1 (Continued)

	GLP Component	Application to an *in vitro* ICS assay
		3. Maintain time-related graphical records of instrument performance (these are built in to most instrument software) 4. Stop using an instrument when performance is outside any acceptable range, and reject any data produced. 5. Instrument must be re-certified after any repair 6. Not included in the guidelines is equipment validation, however, it is critical to maintain all incubators, pipetors, centrifuges, refrigerators, freezers, balances, sterile hoods, and water systems, within expected criteria.
Quality Assurance	Reagent/Material Certification	1. Reagent/materials certification must follow accepted procedures and must be adequately documented. 2. Some guidelines go on to specify that each reagents/materials container must be labeled with its certification value, date, and expiration time, and that expired product must be kept seperate from in-date products. 3. This policy is meant to assure that reagents used are as specified in the SOP's.
	Analyst (Technician) Certification	1. Satisfactory training and/or competence with specific laboratory procedures must be documented such that they are acceptable to any agency or client obtaining results. 2. Criteria are usually prescribed by individual labs, or perhaps by the clinical trial study protocol.
	Lab Facilities Certification	1. Certification of laboratory facilities is normally performed by an external agency for such issues as space (amount, quality and relevance), ventilation, equipment, storage, hygiene, etc. 3. Other certifications are necessary for Safety and GLP.
	Sample/Specimen Tracking	1. Invest in 21 CFR Part 11 -compliant software when available e.g., bar-coded specimens to data retrieval.

	2. Insist on automated Sample ID capabilities in analytical instruments.
Documentation and Maintenance of Records	1. Records may be required in the event of legal challenges due to repercussions of decisions based on the original analytical results. 2. All records must be maintained that pertain to specimen/sample origins, tracking, raw analytical data, processed analytical data, SOPs, instrument validation results, reagent certification results, analyst certification documents, lab certification documents, etc. 3. Traditional Scientist Laboratory Notebooks must be bound, numbered, signed, dated, witnessed, and stored. 4. Keep only records which are specified in approved documents 5. Five years is the general guideline to maintain records.
Accountability	1. Accountability for laboratory results is inherently established when following GLP procedures. 2. Every analyst, instrument, reagent, and analytical method leading to technical results and conclusions must be clearly defined and documented. 3. Allows for quick identification of any source(s) of error(s) to maintain acceptable data quality

Standardization of an assay is an iterative testing process. Empirical or test-derived criteria and ranges must be established and incorporated into the testing process (Colburn, 2003). Cellular functional assays have multiple sources of variability not encountered in other analytical methods. There are several bioassays for measuring cytokine production that range from a single-plex enzyme linked immunoassay (ELISA) capture platform for bulk secreted soluble cytokines, to single parameter indirect single cell secretion measurements (ELISPOT), to multi-parameter cellular assays that either fix the cells (ICS) or that maintains cell viability (Gel Microdrop by One Cell Systems, and MACS™ Secretion Assay by Miltenyi).

Figure 8.1 represents most sources of variability associated with cellular assays. When designing a method, one must consider that each of these will have impact on the consistency of results such that it may prevent

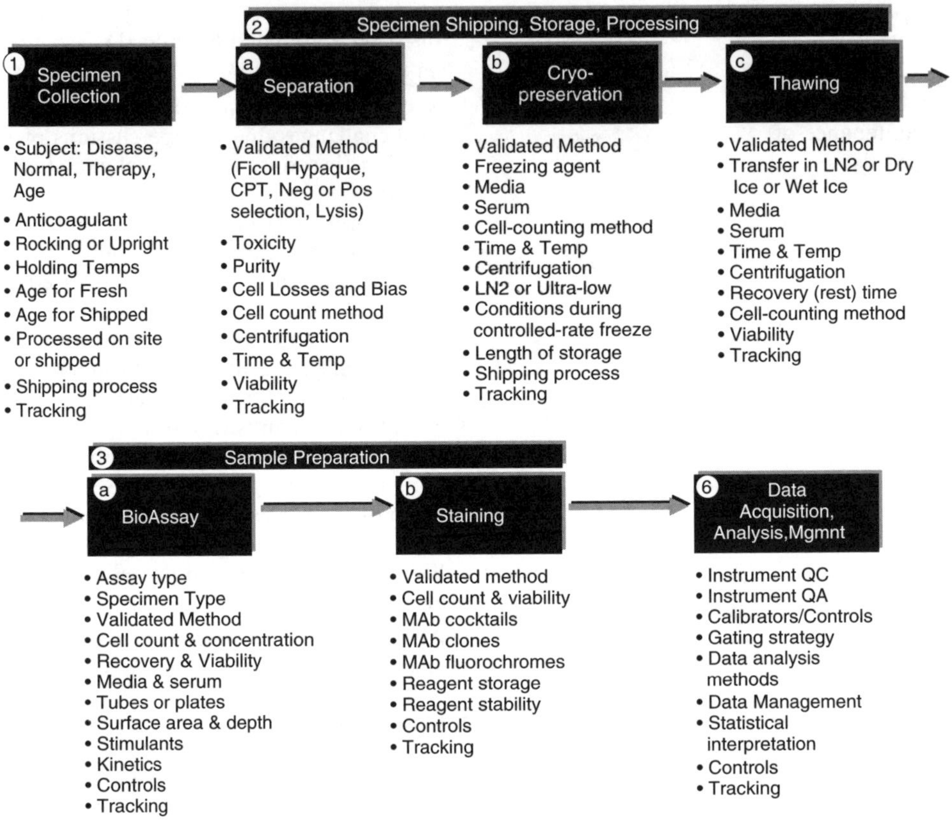

Figure 8.1 Sources of variability

validation of the method. The researcher is responsible for maintaining consistency throughout the development and feasibility process by requiring the same reagent lots, same specimen site, same testing site, etc. and to assign criteria to each so that any deviation is documented and might be used to exclude data from the method validation effort. Examples of hidden sources of variability include false positive results arising from cell culturing with different lot numbers, or different suppliers, of the "same" serum.

The assay test system can be broken down into the critical steps, processes, and desired results (e.g. Table 8.2 for ICS). Then components of each critical step must be defined. For example, the fixation of cell membrane comprises the source of the reagent used, cell concentration, incubation time, and temperature. The same applies for permeabilization and lysing reagents.

Table 8.3 shows a variability checklist designed specifically for ICS. For each variable, specify how it will be controlled (e.g. always use sodium

Table 8.2 Critical steps of ICS

	Intracellular Cytokine Staining Assay
End Result	Reproducible enumeration of rare cytokine-producing T cells from in vitro assay of human whole blood or isolated PBMC (fresh or frozen).
Critical Processes	Fixation of cell membrane Permeabilization of cell membrane Lysing of RBC if whole blood Detection of specific antigens
Critical Parameters	Time Temperature Concentration
Critical Results	Percentage non-debris events Percentage of lymphocytes and subsets Number, Percent, and MFI of positive events Limit of Detection above background/noise

heparin as the blood anticoagulant) for the duration of assay design and validation testing.

8.3 Reagents/controls

Reagents

During assay validation it is advisable to use commercially available reagents and kits that are supplied with detailed method sheets, package inserts, and quality assurance and quality control. Leading reagent providers are ISO9001-certified (International Organization for Standardization) and thus have a Quality System, which define the processes needed to consistently provide products that meet customer and applicable statutory or regulatory requirements.

Fixation and permeabilizing reagents

Several types of fixation and permeabilization reagents are used which may result in differential effects on cells and antigens. Fixatives can affect antigenic sites at certain doses and must therefore be evaluated carefully. Saponin is a reversible permeabilizing agent, and requires stringent conditions and buffers for optimum performance. Knowing that saponin must always be present with the intracellular antigen antibodies, establishing the effectiveness of every lot compared to the current lot must be evaluated. Further, the buffer must not contain added protein, and must be controlled at room temperature, to save lost time from confounding results. The source of water used in all experimentation is another critical factor.

Table 8.3 Variable checklist

Checklist		Variable	Answer
Specimen Collection		·Subject: Disease, Normal, Therapy, Age	
		·Anticoagulant	
		·Rocking or Upright	
		·Holding Temps	
		·Age for Fresh	
		·Age for Shipped	
		·Processed on site or shipped	
		·Shipping process	
		·Tracking	
Specimen Shipping, Storage, and Processing	Separation	·Validated Method (Ficoll-Hypaque, CPT, Neg or Pos selection, Lysis)	
		·Toxicity	
		·Purity	
		·Cell Losses and Bias	
		·Cell count method	
		·Centrifugation	
		·Time & Temp	
		·Viability	
		·Tracking	
	Cryopreservation	·Validated Method	
		·Freezing agent	
		·Media	
		·Serum	
		·Cell-counting method	
		·Time & Temp	
		·Centrifugation	
		·LN2 or Ultra-low	
		·Conditions during controlled-rate freeze	
		·Length of storage	
		·Shipping process	
		·Tracking	
	Thawing	·Validated Method	
		·Transfer in LN2 or Dry Ice or Wet Ice	
		·Media	
		·Serum	
		·Time & Temp	
		·Centrifugation	
		·Recovery (rest) time	
		·Cell-counting method	
		·Viability	
		·Tracking	
Sample preparation	Bioassay	·Assay type	
		·Specimen Type	
		·Validated Method	
		·Cell count & concentration	
		·Recovery & Viability	
		·Media & serum	
		·Tubes or plates	
		·Surface area & depth	
		·Stimulants	
		·Kinetics	
		·Controls	
		·Tracking	
	Stainbing	·Validated method	
		·Cell count & viability	
		·MAb cocktails	
		·MAb clones	
		·MAb fluorochromes	
		·Reagent storage	
		·Reagent stability	
		·Controls	
		·Tracking	
Data acquisition, analyusis, and management		·Instrument QC	
		·Instrument QA	
		·Calibrators/Controls	
		·Gating strategy	
		·Data analysis methods	
		·Data Management	
		·Statistical interpretation	
		·Controls	
		·Tracking	

Wash buffers

Be aware of the composition of your wash buffers related to the assay performance. For example, recipes for PBS vary significantly as do the inclusion and/or concentration of sodium azide and proteins in wash buffers.

Monoclonal antibodies

The implementation of the cluster designation (CD) nomenclature for antigens and monoclonal antibodies, along with the commercial availability of pre-formulated multi-color reagents, have reduced the awareness of the dependence of assay results on the choices of staining reagents.

Creating reproducible multi-color cocktails is based upon the understanding of the consequences related to choices made for clone, affinity, avidity, isotype class, dose, and the source, type and size of the fluorochromes used for each MAb (monoclonal antibody) in the mixture, along with the instrument and optical filter choices.

Commercially available multi-color products with defined stabilities are ideal for use in an assay due to rigorous product development; however, it is difficult to find exactly what you need. Be aware that labs may perform custom mixing of single antibody reagents into cocktails without performing appropriate optimization, validation, and stability testing of the new combination. The next best thing to a multi-color catalog product is a manufacturer's custom-formulated cocktail which has been designed, optimized, and tested by dedicated experts according to the user's assay requirements. Although these custom cocktails can seem costly up front, they can prove to be very cost effective when weighed against erroneous results.

Understanding lyophilization

Lyophilization gives unstable chemical and protein solutions a longer shelf life at room temperature and excellent solubility for rapid reconstitution. Over the years, reagent and kit manufacturers have explored alternative types of containers for the processing and delivery of lyophilized (freeze-dried) reagents and have not had great success with any vials other than the traditional glass vials with slotted rubber septa. The physics behind lyophilization imposes rigid vessel requirements essential to the process, and improper lyophilization will greatly compromise the integrity and stability of the product. Always verify the claims of lyophilized products by reviewing GMP manufacturing records and specific performance testing. If performance testing is not available, then it is up to you to perform these studies prior to incorporation into your experiments.

When using a lyophilized reagent in an assay, three points must be considered: is the product high quality freeze-dried or just vacuum-dried, what is the real time shelf-life and storage condition, and how does the lyophilization process affect the activity of the chemicals. Calibrators are required to validate instrument performance and are often suspensions of microscopic beads. The package inserts will describe the proper method to re-suspend beads and maintain them in suspension.

Controls

For ICS, consistent negative and positive controls must be in place and should be specified in the study-specific protocol. The raw data obtained for controls should be archived along with the study results.

Assay controls

Assay controls are designed to detect any changes in the entire test system in either direction of response, and are included in every assay to evaluate the performance at a given time and over the duration of the study. They can be a combination of multiple step-specific controls, or a single control designed to do the whole job. A single control designed to test that all portions of a test system are functioning optimally will only tell you when the system succeeded or failed. While this is the GLP requirement, it won't help the investigator determine the root cause of a failure so that corrective and preventative actions (CAPA) can be taken.

All manipulations specified for the test system protocol must be carried out on the controls. Once established, all control ranges should be re-verified when a major change is introduced to the assay system (such as a new type of culture media). New lots of controls need to be tested concurrently with the previous lot until proven equivalent or new ranges have been established. The ICS assay test system has five segments to control: the specimen type, the *in vitro* culture, the staining method, data acquisition, and data analysis.

An example of a good control for the entire ICS assay system of cryopreserved PBMCs, is a large batch of frozen vials of PBMCs from a known positive responder, which will be thawed and cultured with and without stimulation, and stained with and without isotype controls for each assay performed. This donor should have established ranges for performance in the assay. A bio-statistician will calculate the number of tests required to determine ranges.

For cell-based assays, many investigators use PBMCs obtained by leukopheresis since it can generate such large numbers of cells from a single donor at a single time. There have not been any definitive studies published which compare the functional status of cells obtained via leukopheresis to cells obtained from venous puncture; therefore, this source of positive control cells is yet another variable to be specified.

A known stimulant is used on the test sample to demonstrate the capacity of test cells to respond to an *in vitro* stimulation (i.e. verification that a negative response is not due to the cells being incapacitated or dead). It is best to use a stimulant that mimics the type measured in the test specimens (e.g. a pool of common peptides is not a good positive control for an intact protein test stimulant which requires antigen processing by the cells). There are many other well-characterized T-cell stimulants to choose from such as Staphylococcus Enterotoxin-B (SEB), Phorbol 12-Myristate 13-Acetate (PMA), Ionomycin, Phytohemagglutinin (PHA), Concanavalin-A, recall antigens, viral vectors, or anti-CD3+CD28 to name some.

Additionally, the whole blood cell count and differential should be measured just prior to beginning the assay. For PBMCs, measure the cell count and viability, and if cells were cryopreserved, calculate the percentage of viable cells recovered from the total number of cells thawed. PBMC samples should also be inspected under a microscope. All the values must be recorded and stored with the data.

In vitro *culture bio-assay controls*

During *in vitro* stimulation there are positive and negative controls for the complete culture conditions (media/serum, incubator, and the stimulants), therefore, cells known to reproducibly respond to the test stimulant, or at least to a positive control stimulant, need to be characterized and used during every assay. Ideally the same cell source is used for controls and test specimen (e.g. venous puncture whole blood in sodium heparin, or cryopreserved PBMC) and the positive control should respond at comparable levels compared to the test specimens.

Finally, a negative control is needed to monitor non-specific and background stimulation in the bioassay and can be as simple as the test cells alone in culture media during the culture period. Alternatively, you can include a mock stimulant (e.g. irrelevant peptides or stimulant solvent alone) into the negative cultures of the positive control cells and/or test cells.

Staining method controls

This requires a consistent source of control cells with a defined and reproducible range of negative and positive responses to be measured, and that are equivalent to the test samples. These can either be the assay test system control cells, pre-activated and stored cells, commercially available control cell products, or cell lines expressing the antigens. The staining method comprises the addition and integrity of antibodies, fixation, permeabilization, and washing processes.

Isotype antibodies are used to control for non-specific and background level binding of the antibody protein and fluorochrome if directly conjugated. Since every monoclonal antibody is a unique protein, this binding is at best, an estimation. There are differing opinions on this point, so the user must assign a level of significance to isotype measurements. It is critical when measuring intracellular antigens to use some type of control for the non-specific interactions due to the highly complex milieu present in the cytoplasm and the charged nature of proteins. Isotype antibodies as negative controls are very useful for setting the analysis regions reproducibly and this is especially useful for heterogeneous activation markers like CD69 and IFNγ, and others such as CD45RA.

Data acquisition controls

All equipment must be certified, calibrated, validated, and routinely tested for performance. Instruments must be validated meticulously, and tested conscientiously. All instrument systems used to acquire data must share the same performance characteristics. Each should perform the same over time, otherwise the data most certainly cannot be compared between days and instruments, and the data integrity may be called to question.

The basic validation of daily instrument performance is completed using commercially available calibrators which are generally polystyrene beads of proven size and homogeneous distribution with assigned levels and ranges of fluorescence intensity. Some are designed for specific instrument manufacturers and models, but most can be used across platforms. Fluorescent target channels for intensity are critical to multi-instrument, multi-site and longitudinal comparisons of results. Each assay method must provide for determining the specific target channels and their subsequent use. Every day of sample acquisition, the detectors must be verified using beads with defined fluorescent intensities.

There are several methods available to validate performance. The main resource for information about the qualification, validation, performance testing methods, and record keeping, is the manual accompanying each instrument, or other manufacturer-supplied documentation.

Electronic compensation of the spectral overlap between fluorochromes (color compensation) is difficult to standardize to a degree that no post-acquisition analysis is required. There is good automated software available for compensation calculations from major instrument manufacturers, and these should be used; however, the material used to determine compensation values can cause problems. For example, some labs use dye-loaded beads, or a commercial control cell, or the actual test specimen to establish color compensation values. The MAb and fluorochromes chosen are also variable. Methods range from a single Ab clone conjugated to each fluorochrome used, to combinations of mutually exclusive markers (e.g. CD4/CD8), to the actual test reagents. Also complicating the picture is the fact that a fluorochrome bound to an intracellular antigen in the test sample may not be detected the same as when it is bound to a surface antigen on the cells or loaded in a bead that was used for compensation.

Data analysis controls

The same set of list mode data files could be used for every data analysis session, user, and for every analysis site to confirm and validate the software performance, gating strategies, analysis region placement, and statistical value calculations. Since the instrument acquisition protocol determines how the data may be analyzed, stored and presented, it is important to acquire the same parameters in the same manner across instruments within a lab or across many labs (e.g. the inclusion of time as a parameter).

Standard methods of event gating and population analysis must be established for each assay and possibly each tube (e.g. isotype, non-stimulated, stimulated). Data report templates should be designed for implementation across instrument platforms if required. This can be accomplished by using known list modes to validate any new or modified analysis protocols or reports, and to share this list mode between all sites.

8.4 Breaking down the ICS assay method

Key publications describing intracellular cytokine assays include the original paper by Jung et al (1993), which was improved further by Picker et al (1995) Currently there are many variations on the "fix and perm" theme such as: the lysis of RBC may not be included, or can occur during fixation or permeabilization. Also, when the cell surface antigens are known to survive the fixation treatment, all of the antibodies can be added during or after the permeabilization step.

All intracellular assays have two basic steps: stabilization of the cell membrane to maintain cell integrity and then permeabilization of the membrane for access to cytoplasmic proteins. Nuclear proteins generally require additional treatments that may preclude simultaneous measurement of surface and cytoplasmic proteins, and signal transduction pathway molecules require different fixation methods to immediately stop a phosphorylation action; however, these methods will not be discussed here.

Figure 8.2 illustrates a manually performed ICS method. The specimen and surface antigen antibodies are incubated together; the sample is then

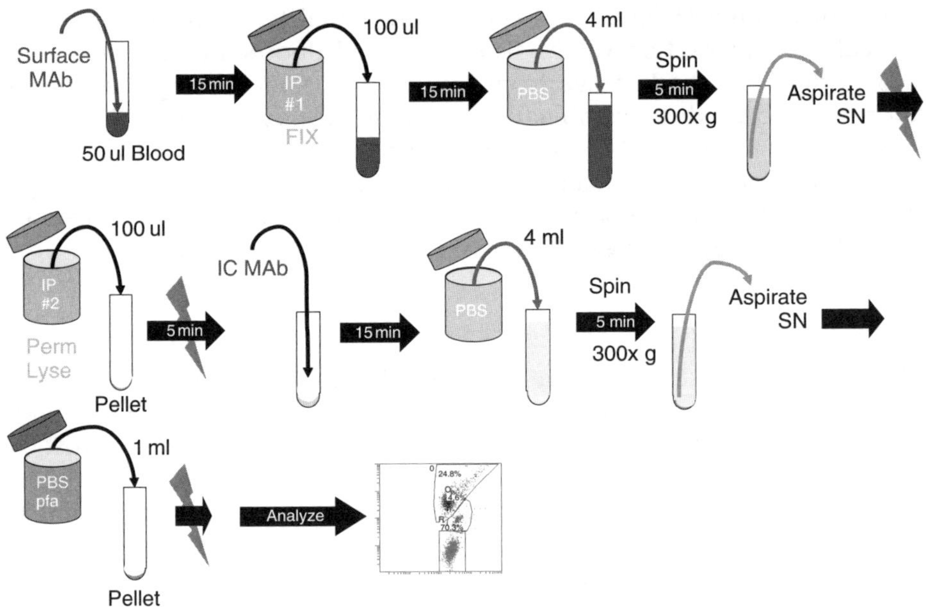

Figure 8.2 Manually performed ICS method

fixed and washed, a cell membrane permeabilization agent is added along with intracellular antigen antibodies, the sample is washed again prior to analysis on a flow cytometer. In this example, IP refers to IntraPrep reagents from Beckman Coulter, Inc., and pfa is paraformaldehyde.

Determine the constants

Of all the potential variables known in ICS we defined our response to each of the line items in Table 8.3. This ranged from a decision to keep a variable constant, to simply recording the information during each assay test for later analysis of assay performance.

Design the controls

All negative and positive controls are identified for the ICS assay in Table 8.4 and were incorporated into the assay design for use with a model system to work out the assay methods.

Automatable method

The automated ICS staining method shown in Figure 8.3 was refined and optimized for testing cell surface CD19 and intracellular cytoplasmic

Table 8.4 The controls for ICS

Critical Steps	Controls	Examples
Assay System Control	Negative	Frozen donor defined performance ranges
	Positive	Frozen donor defined performance ranges
Sample Preparation	Negative	n/a
	Positive	Test Sample stimulated with known positive stimulant, AND Frozen donor defined performance ranges, or Heparin whole blood from pre-characterized normal donor
Sample Culture	Negative	No stimulants added to: Test Sample alone in culture (non-treated), AND Frozen donor defined performance ranges, or Heparin whole blood from pre-characterized normal donor
	Positive	Defined Stimulants SEB+CD28 added to: Test Sample, AND Frozen donor defined performance ranges, or Heparin whole blood from pre-characterized normal donor
Staining Method	Negative	Isotype Ab Cocktail on: Stimulated test specimen, AND Non-stimulated Frozen donor defined performance ranges, or Heparin whole blood from pre-characterized normal donor
		Stimulated Antigens Ab Cocktail: Non-stimulated test specimen, AND Stimulated Frozen donor defined performance ranges, or Heparin whole blood from pre-characterized normal donor
	Positive	Frozen donor defined performance ranges, or Heparin whole blood from pre-characterized normal donor
Data Acquisition	Negative	Cell suspension buffer sans cells
	Positive	Instrument Fluidics, Optics, and Alignment Beads Instrument Detector Optimization beads (i.e. target channel) Color Compensation using CD45 conjugates of each color using: Frozen donor defined performance ranges, or Heparin whole blood from pre-characterized normal donor
Daya analysis	Negative & Positive	Frozen donor defined performance ranges, or Listmode file with known data stored from validation testing

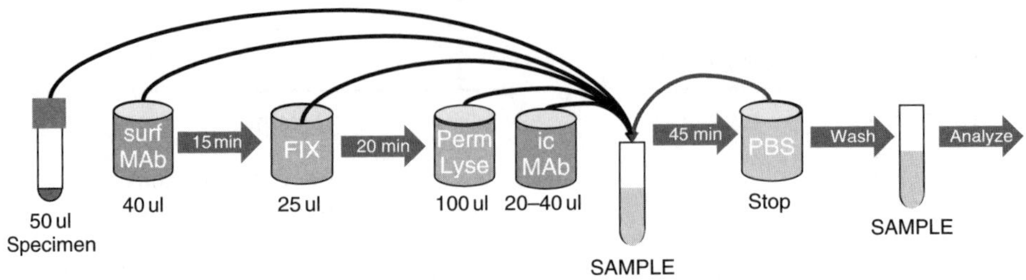

Figure 8.3 The automated ICS assay design

CD79a. The benefits of this model were that all normal human blood contains large numbers of B cells (CD19+) and each B cell constitutively expresses CD79a protein in the cytoplasm. Therefore, for the purposes of staining method design, the variables associated with sample preparation and in vitro culturing were removed. For the entire assay system performance and validation testing, the intracellular cytokine IFNγ was measured in normal human T cell subsets expressing CD69 from in vitro stimulated whole blood or PBMC cultures.

8.5 Assay reagents, controls, acquisition and analysis

Only commercially available reagent formulations were used throughout assay design and testing. Each flow cytometer used in testing was completely validated for equivalent performance and tested daily for fluidic system function, optical alignment and detector system performance. Automated color compensation was performed daily using commercially available lyophilized lymphocyte control cells stained with CD45 antibodies and washed once. This method gave comparable results to the test specimen prepared via the automated ICS method with CD45 antibodies.

Lymphocytes were gated automatically based on forward and side angle light scatter, and all analysis regions were set automatically by the cytometer acquisition software to include no more than 0.5% of positive events in the isotype control tube of non-stimulated culture cells. The background level of activation was measured from the non-stimulated test culture stained with the activation antibodies, and was subtracted directly from the stimulated test culture values.

Performance testing of the automated ICS assay demonstrated an accurate, precise, linear, specific, and sensitive method. The Limit of Quantitation (LOQ) was not the same in CD4 and CD8 populations of T cells. CD8+ cells had an LOQ of 0.24%, and CD4+ was 0.16%. However, these

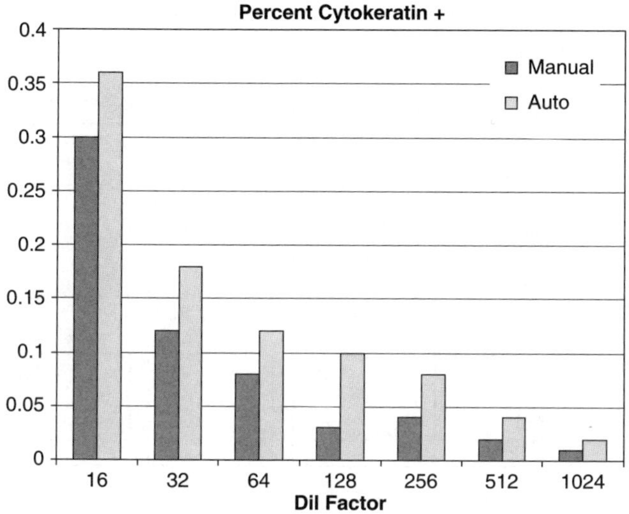

Figure 8.4 Linearity testing

calculations were based on traditional analytical method validation testing which does not take into account the clinical relevance of bioassay data. A biostatistician can help create a more relevant clinical value for cut off criteria. Intra-assay precision was compared to the traditional manually performed ICS method and with a five-color flow cytometry assay to define the population of interest (e.g. CD3+CD4+CD8–CD69+IFNγ+), the %CV was consistently <9% with mean values of 1–4% positive. Inter-assay precision (and specificity) was determined to be 14% using the automated ICS method on greater than 75 specimens with mean positive values <0.5%.

Colo-205 is a colon adenocarcinoma epithelial cell line that expresses intra-cytoplasmic cytokeratin and cell surface CD71. When these cells are diluted into human whole blood and prepared for flow cytometry, they are distinguished easily from WBC populations based on light scattering properties. The linearity testing of the automated ICS method was performed with Colo-205 serially diluted into normal human whole blood. A representative example of a dilution at the lower end of detection is shown in Figure 8.4.

8.6 Results

The following figures are shown to demonstrate the dramatic impact on standardization that automation brings to functional cellular analysis. The

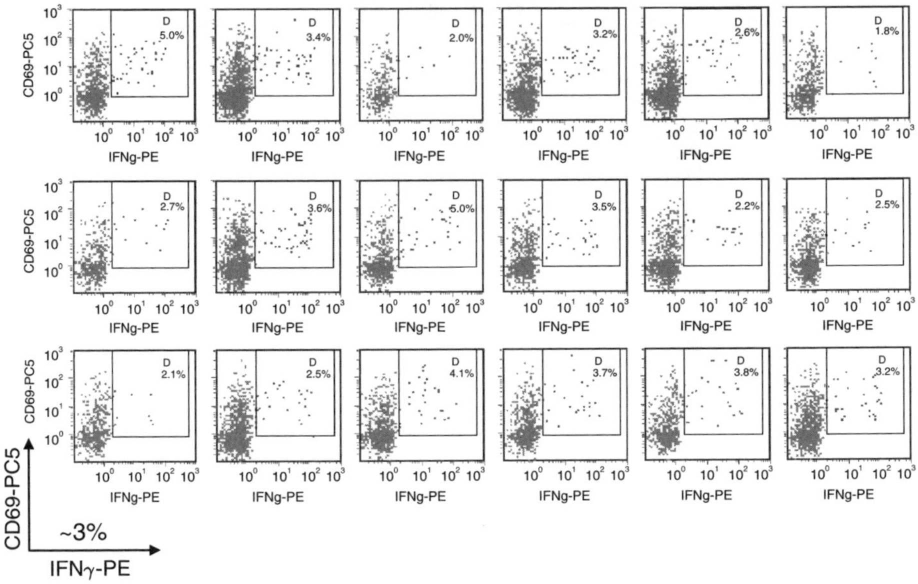

Figure 8.5 Manual ICS assay method replicates: CD4+ lymphocytes from SEB+CD28 stimulated whole blood

same stimulated blood specimens were stained on the same day for IFNγ expression using both the traditional manual ICS method (Figure 8.5), and an automated ICS method (Figure 8.6). Because this is an automated method, there are fewer assay ranges that must be established and validated. For example the assay designer does not need to establish ranges for incubation times, reagent volumes, and centrifugation speeds. Additionally there is no longer the need to certify reagent additions since most liquid handlers have liquid sensing capability and notify the user when a vial of reagent is low in volume.

8.7 Conclusion

Standardization of functional cellular assays acceptable for use in multi-site clinical trials and biomarker discovery programs is possible through strict adherence to good laboratory practice guidelines throughout the design, optimization, validation, and performance. The time and resource investment to ensure GLP adherence ensures consistent and accurate results which in turn helps interpret data more meaningfully.

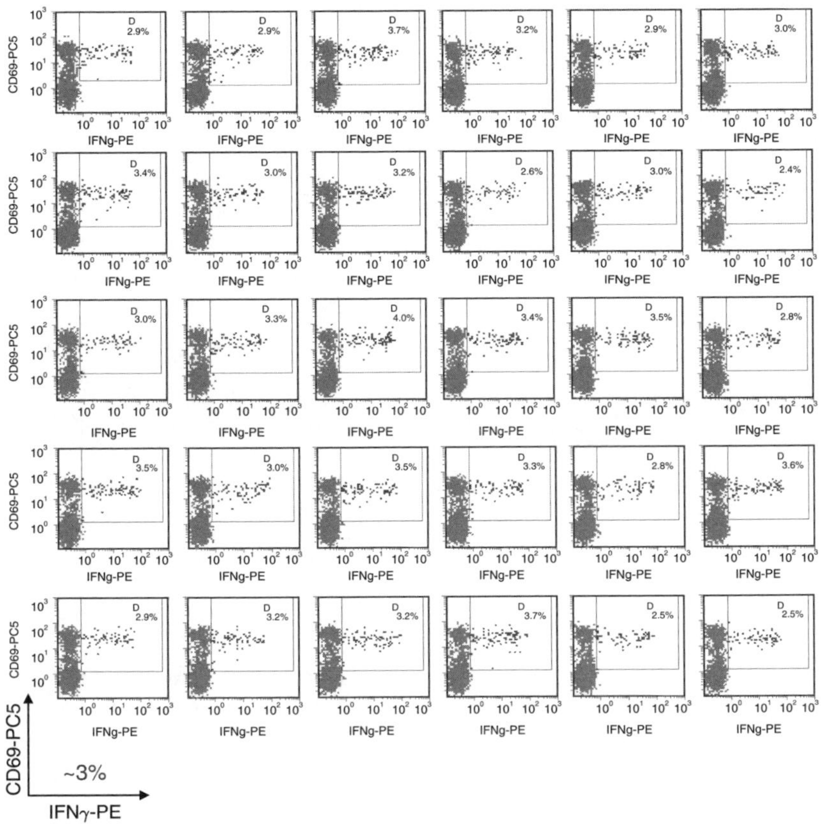

Figure 8.6 Automated ICS assay method replicates: CD4+ lymphocytes from whole blood stimulated with SEB+CD28

References

Cox JH, G. Ferrari, SA Kalams, W Lopaczynski, N Oden (2005). Results of an ELISPOT proficiency panel conducted in 11 laboratories participating in international human immunodeficiency virus type 1 vaccine trials. *AIDS Res Human Retroviruses*, **21**, 68–81.

Maecker HT, A Rinfret, P D'Souza, J Darden, E Roig (2005). Standardization of cytokine flow cytometry assays. *BMC Immunol*, **24**, 6–13 doi:10.1186/1471-2172-6-13.

Colburn WA (2003). Biomarkers in drug discovery and development: from target identification through drug marketing. *J Clin Pharm*, **43**, 329–341.

Jung T, U Schaver, C Heusser, C Neumann, C Rieger (1993). Detection of intracellular cytokines by flow cytometry. *J Immunol Methods*, **26**, 197–207.

Picker LJ, MK Singh, Z Zolrakeski, JR Treer, SL (1995). Direct demonstration of cytokine synthesis heterogeneity among human memory/effector T cells by flow cytometry. *Blood*, **86**, 1408–1419.

9 Validating reference samples for comparison in a regulated ELISPOT assay

Magdalena Tary-Lehmann[1,2]*, Christina D. Hamm[1] and Paul V. Lehmann[1,2]

[1]*Cellular Technology Limited*

[2]*Department of Pathology, Case Western Reserve University, Cleveland, OH 44106, USA*

9.1 Introduction

In order for pharmaceutical and biotechnology companies to market their products in the United States or abroad, they must first receive approval from the regulatory /governing agency in the country they reside, i.e.,

*Address correspondence to: Magdalena Tary-Lehmann, Cellular Technology Ltd, 10515 Carnegie Ave., Cleveland OH 44106, Phone: (216)791-5084. Fax: (216)791-8814. E-mail: mtl@immunospot.com and or mxt27@po.cwru.edu

United States Food and Drug Administration (US FDA), Organization for Economic Co-operation and Development (OECD), European Union (EU), etc. Companies are required to determine immunotoxicity and immunogenecity for their product before they can submit their application (Immunotoxicology Studies for Human Pharmaceuticals, ICH Guidance Document Status: Step 5 September 2005). Guidance documents from the US FDA describe immunology assessment requirements for products headed towards an Investigative New Drug Application (IND) submission (US FDA Guidance Document Immunotoxicology Evaluation of Investigational New Drugs October 2002). In order to receive approval from one of the regulatory agencies, the company must manufacture and test their products according to the regulations stipulated by their government agency. If the company wishes to market their product in another country, they must submit an application to the regulatory agency which governs in the country of interest.

With this understanding, both the manufacturer and any contributing laboratories performing regulated testing must have a sound quality system in place, ready to support the regulatory requirements of their submission (GLP Quality Audit Manual, 2nd edition, Milton Anderson, 1992 and FDA Good Laboratory Practice Regulations (21 CFR Part 58). Development of a quality system starts with researching the regulatory requirements and developing a quality plan. The plan will outline the resources necessary for a testing facility to conduct regulated studies, i.e., facilities, equipment, personnel, written standard operating procedures (SOPs), training programs, etc. Such a quality system is defined and maintained by two main groups: Quality Assurance (QA) and Quality Control (QC). In a Good Laboratory Practice (GLP) regulated environment (Good Laboratory Practice Regulations, 2nd edition, Sandy Weinberg (ed.), 1995. Good Laboratory Practice Standards, U.S. EPA, Office of Prevention, Pesticides, and Toxic Subsohn Robertson, Virginia-Maryland Regional College of Veterinary Medicine, 1991)) QA is responsible for inspecting regulated studies, submitting periodic reports to Testing Facility Management, and remaining *independent* of study conduct, thereby providing an *Outsider's Perspective*. Conversely, QC is not required in a GLP regulated environment, but is important for ensuring data integrity, or an *Insider's Perspective*. QC can be as simple as one scientist signing off on another scientists' data after review, or it could be as advanced as 100% check of electronic data transfer, data point by data point. A quality system must be clearly defined and communicated to all who work with the system.

After the establishment of a quality system, the performing of a regulated study requires that the testing facility validate the test method prior to running the test samples. When applied to immunological assessment, such validation is dependent on many factors and is defined by the readout system to be used. Several potential alternatives include enzyme-linked immunosorbent assay (ELISA), enzyme-linked immunospot (ELISPOT) and flow cytometric methods fluorescence-activated cell-sorting (FACS). An examination of ELISPOT and how it compares to these alternatives aids in exemplifying the need to validate this most useful assay.

9.2 Defining ELISPOT

Every T cell expresses a unique T cell receptor (TCR), which is specific for a single antigen. In order to be able to recognize the infinite numbers of potential microbial antigens, the T cell system relies on an almost infinite number ($\sim 10^{12}$) of different possible T cell specificities. Subsequently, the frequencies of naive T cells recognizing individual antigens are very low. As a result of the clonal expansion associated with infection, the frequency of antigen-specific memory/effector cells can transiently rise to as high as 1:100 among white blood cells. However, for long term memory cells and in chronic infections these frequencies typically settle in the range of 1:10 000 and much lower. The frequencies of specific T cells in the blood after immunizations with subunit vaccines (proteins/peptides) and environmental sensitization also rarely exceed the 1:10 000 threshold. Therefore, techniques that measure T cell immunity need to be sensitive enough to reliably detect the rare antigen-specific T cells at the 1:10 000 threshold or below. In addition to measuring the frequencies of the antigen specific T cells, which delineate the magnitude of immunity, it is also essential to understand the quality of the T cell immunity. It depends on whether the T cells are Th1 or Th2 type (for CD4 cells) or Tc1/Tc2 for CD8 cells, and whether they are capable of direct killing. Sophisticated assays therefore will also provide information on the effector lineage of the antigen specific T cells. In ELISPOT assays, the production of cytokine (Czerkinsky et al, 1988) or any other released effector molecule (for example granzyme B or antibodies) can be measured directly *ex vivo* (Czerkinsky et al, 1983; Rininsland et al, 2000). Cells (peripheral blood mononuclear cells [PBMC] generally or purified cell populations), are plated in the absence of antigen and the constitutive production of secreted molecule forms the background against which antigen induced responses are compared. Experimental wells contain test antigens. The presence of the antigen/"physiologic" stimulation induces the production of cytokine (or any secreted molecule). As these molecules are secreted by the individual cells within 4–48 hours they are captured in the proximity of the secreting cell by capture antibodies on the plate surface (Figure 9.1). This short assay can detect *in vivo*-differentiated memory T cells.

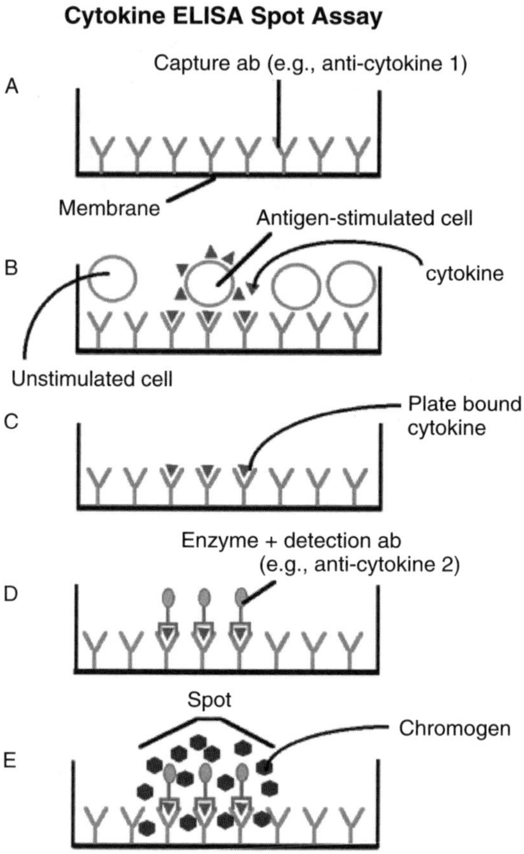

Figure 9.1 Schematic diagram of ELISPOT assay. 96 well plates are coated with capture antibodies (A). Responder cells are added to the wells with or without antigenic stimulus. As the individual cells respond and secrete cytokines or other products, the products are bound by the antibody directly at the site of secretion (B). Cells are washed off (C), second detection antibodies are added (D) and the spots are developed with chromogen (E). Each spot represents cytokine produced by a single cell

9.3 Comparison of the performance and utility of ELISPOT assays with other T cell diagnostic techniques

Cytokine ELISA assays, *cytokine bead arrays (CBA)* and *cytokine protein arrays (CPA)* measure the bulk cytokine that the cells release into the culture supernatant. These assays have the following fundamental disadvantages relative to ELISPOT.

Sensitivity: Cytokine measurements on culture supernatants are at least 200 times less sensitive (Tanguay and Killion 1994) in

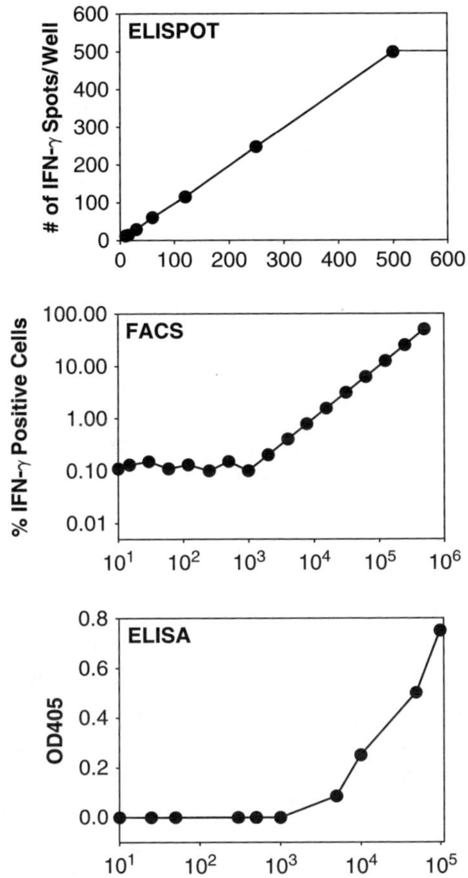

Figure 9.2 Unique sensitivity of the ELISPOT assay for the detecting low frequency cytokine producing cells. The specified number of IFN-γ producing T cells was mixed into 1 million antigen presenting cells. The number of IFN-γ producing cells was measured using ELISPOT (upper panel) and intracytoplasmatic staining (middle panel); result of ELISA measurements are shown in the lower panel

detecting low frequency antigen specific T cells than is ELISPOT (Figure 9.2). The sensitivity of ELISPOT assays has been dramatically increased by the introduction of polyvinylidinene fluoride (PVDF) membranes (Schielen et al, 1995). The use of PVDF membranes has been, along with computerized image analysis, the major steps in making ELISPOT the robust, high resolution assay that it is today (Anthony and Lehmann 2003; Carter and Swain 1997; Forsthuber et al, 1996; Heeger et al, 1999; Helms et al, 2000; McCutcheon et al, 1997). The reason for ELISPOT's increased sensitivity in comparison with *ELISA*-type assays is that in ELISPOT assays the cytokine secreting cell directly sits on the membrane and the cytokine is captured around the secreting cell as it is released where the local concentration of the cytokine is high (Figure 9.1). In ELISPOT assays, the cytokine is captured before it becomes diluted in the supernatant, captured from the supernatant by (abundant) cytokine receptor-bearing bystander cells, and before it is degraded by enzymatic cleavage. *ELISA* , *CBA* and *CPA* detect cytokine in the supernatant, after dilution, absorption, and degradation has already occurred and thereby explaining the ELISPOT assay's much higher sensitivity when it comes to detecting low frequency cytokine secreting cells (Sedgwick and Holt 1983). T cells specific for recall antigens such as tetanus toxoid, candida, measles-, mumps-, flu-, cytomegalo- and Epstein–Barr virus typically occur in the low frequency range below 1:10 000 antigen-specific T cells in PBMC (Kreher et al, 2003). Also T cells that recognize HIV and hepatitis C peptides are commonly less than 1:10 000 in PBMC (Anthony et al, 2002; Kleen et al, 2004). The ELISPOT assay has, in principle, no inherent lowest detection limit and can be configured to measure as few as 1 in 100 000 000 cytokine producing cells.

Per cell cytokine productivity. In the ELISPOT assay the size of the spots adds additional information to spot number. Spot size reflects how much cytokine or released product the individual cells produce. Cytokine protein measurements done on culture supernatants (*ELISA*, *CBA*, *CPA*) or *cytokine mRNA determinations (RT-PCR)* do not account for the productivity of individual cell and hence do not permit to distinguish whether few cells produce much cytokine or many cells produce little cytokine. The amount of cytokine a T cell produces depends on the extent of T cell stimulation. At a given antigen dose, high avidity T cells produce considerably more cytokine than low avidity cells do. At the same time cytokine productivity is affected by the time period that has elapsed since the T cell's last antigen encounter (Hesse et al, 2001). It also has been shown that the per cell cytokine productivity of T cells is highly variable over a wide range. Therefore, employing ELISPOT assays allow the accurate clonal sizes along with the per cell productivity to be revealed.

Resolution: Because ELISPOT assays detect individual cytokine secreting cells, the data obtained in replicate wells can be subject to stringent statistical analysis providing clear-cut results even in low signal to noise scenarios.

ELISPOT, *tetramers/pentamers* and *intracytoplasmic cytokine staining (ICS)* are single cell assays that permit the numbers of antigen specific T cells to be counted in cell material such as blood. The accurate frequency information provided translates into the magnitude of antigen specific T cell immunity. Defining frequencies of the specific T cells is one of the primary goals of *ex vivo* T cells diagnostician.

ELISPOT versus tetramer and pentamers which are multimerized MHC molecules

Tetramers and *pentamers* are multimerized MHC molecules. When loaded with a specific peptide, they represent the T cell receptor (TCR) ligand for peptide specific T cells. Specific T cells can be stained with *tetramers/pentamers* and subsequently detected by *flow cytometry*. For reason of feasibility, tetramer and pentamer studies are required to be limited to individuals that express a common allele, such as HLA-A2. This shortcut excludes all non-allele positive individuals from the study and omits all the other HLA gene products expressed in that individual rendering the immune diagnostic information gained incomplete. Each of the six HLA Class I alleles can be equally important for restricting antigen specific CD8 cells, A2 not being more important than any of the other restriction elements. Comprehensive *tetramer/pentamer* analysis would therefore require complete high resolution HLA-typing of every study subject and customization of *tetramer/pentamers* for the all the HLA molecules expressed by the individual. The required effort and cost makes this customization inconceivable for a clinical trial or study. Even if undertaken, this monumental effort would be just the first step leading to an even larger challenge: the choice of the peptide-antigen.

ELISPOT and intracellular cytokine staining (ICS) both follow similar principles. *ICS* measures intracellular cytokine while ELISPOT measures the actually released biologically relevant cytokine since cytokine production can be regulated post-translationally. Cytokine synthesized by a cell but not released is "irrelevant" because it will not exert any effector function *in vivo*. Since the tested cells survive ELISPOT assays, the T cells can be readily propagated subsequently in other platforms or can be frozen for re-testing (Ott et al, 2005). *ICS does not allow for this*. Recognizing that optional maximization of sample resources are critical in clinical trials where the numbers of PBMC available is frequently a limiting factor makes ELISPOT ideal. *ICS*, together with *flow cytometric* measurements excel in measuring cells that occur in relatively high frequencies and can be done post-ELISPOT if needed. Detecting positive cells in lower frequencies than 1:1 000 is difficult (Figure 9.2) (Carter and Swain 1997; Heeger et al, 1999; Helms et al, 2000). The ELISPOT approach permits the measurement of the actual effector function, for example the cytolysis of target cells (Kleen et al, 2004; Rininsland et al, 2000). ELISPOT assays can be done with cryopreserved PBMC without loss of function and provide high inter- and intra-assay reproducibility (Gebauer et al, 2002; Kreher et al, 2003). For all the above reasons, ELISPOT assays have become the technology of choice for human *ex vivo* T cell diagnostic. In recent years they have developed into an indispensable tool for studying infections such as HIV and hepatitis C and for clinical trials that aim at monitoring specific T cell immunity.

9.4 Parameters impacting the ELISPOT assays

Most frequently the 96-well plate format is used in the ELISPOT assay. ELISA plates (Czerkinsky et al, 1983; Sedgwick and Holt 1983) and filter plates using nitrocellulose membranes (Czerkinsky et al, 1988) or PVDF (Forsthuber et al, 1996; McCutcheon et al, 1997; Schielen et al, 1995; Shaw et al, 1993) are used. There are many parameters including technical, biological and reagents variables which can also influence the ELISPOT assay and must be selected during assay optimization. Examples are listed in Table 9.1. Cell preparations used in the ELISPOT should be plated according to viable cell numbers. The kinetics of the secreted product (McHugh et al, 1996) will define the incubation times of the cells with the antigen. Viability of counted cells can be determined for example by acridine orange (AO) ethidium bromide (EB) staining after thawing of frozen PBMC. Only viable cells are capable of performing the complicated mechanism leading to the secretion of the measured product in an ELISPOT. Cryopreserved PBMC can be used (Gebauer et al, 2002; Kreher et al, 2003) after appropriate freezing and thawing. The temperature of the solutions used during the freeze and thaw procedures is vital (Kreher et al, 2003) as well as the media and solutions. If serum is used for the preparation of media it needs to be heat inactivated and screened in biological systems, best in ELISPOT itself, to ensure the absence of mitogenic and toxic effects. Antibody (Ab) pairs used for the ELISPOT assay should be cross-titrated to optimize the assay using an antigen specific read out system (Figure 9.3). The antigen used in the assay should be titrated to find the optimal antigen concentration (Hesse et al, 2001). This will reduce the signal to noise ratio and as well optimize the quality of the spots seen in the assay leading to distinct and easy to count spots. The quality of the spots/assay will improve the accuracy of the counted frequency of the antigen specific cells as well reduce the errors between the replicates. This cross-titration of the antibodies is again needed when a new batch is used. The antigen used in the assay should also be titrated to find the optimal antigen concentration (Hesse et al, 2001). It should be noted that different antigens (Ag) will stimulate different cell populations. Short peptides induce $CD8^+$ T cell responses, whereas longer peptides and proteins will activate the CD4 T cell. Live viruses may activate the innate immune system, NK cells as well as the T cell population. Different mitogens can activate different cell populations (Kleen et al, 2005; Kreher et al, 2003)and could be used as assay controls to ensure that the tested cell population is capable of producing cytokines. To have an antigen specific control of activated memory T cells peptide pools could be used, for example, the CEF peptide pool for human $CD8^+$ cells. This peptide pool covers HLA class I T cell epitopes of cytomegalovirus, Epstein–Barr virus and flu virus and elicits recall responses in most humans (Currier et al, 2002). Different cells populations will secrete different amount of

Table 9.1 Main variables which impact the result of an ELISPOT assay

	Results/Options	Impact/Resolution
Biological variables		
Cell number	TNTC/ no spots countable spot numbers	– titrate the cell numbers to antigen – frequency of antigen specific cells
Incubation time	For example 4 hours GzB 24 hours for IFN-γ, IL-2 48 hours for IL-4, IL-5	– kinetics of secretion of the product
Antigens	Proteins Short peptides Long peptides Live viruses Other	– CD4 T cell responses – CD8 T cell responses – CD4 T cells – $CD8^+$ T cells and/or NK cells – for other cell types
Reagent variables		
Antibodies	ELISA effects, diffuse spots	– cross-titration of antibodies for optimal spot morphology
Plates	defective plates leak problems background noise	– pretest plates
Media	serum free serum containing	optimized for ELISPOT ELISPOT assay screened serum, heat inactivated, need big batches
Substrates	color development time	diverse substrate selections defined by the Ab and substrates used
Environment	incubator, humidity CO_2	dependent on the cells used
Antigen	coloration of membranes	purity of antigens/peptides should be high
	peptide kills cell	solvent should be pre-tested
Technical variables		
Washing	in the center of the wells are	less forceful dispensation of fluids
	bare areas on the membrane	optimizing distance of ELISA washer needles
Incubator shelf	accumulation of spots in all wells and plates unilaterally	evening of shelving or incubator or floor
Pipetting	uneven distribution of spots in the well	pipet cell solutions last use horizontal pipetting method and no angle
Coating solution	defined white areas in wells	avoid bubbles during pipetting check for even coating of the wells
Counting	variable results from experiment to experiment	plate cell number according to viable cells (AO/EB)
Freezing/thawing	not enough viable cells	use approved protocols, no cold Solutions

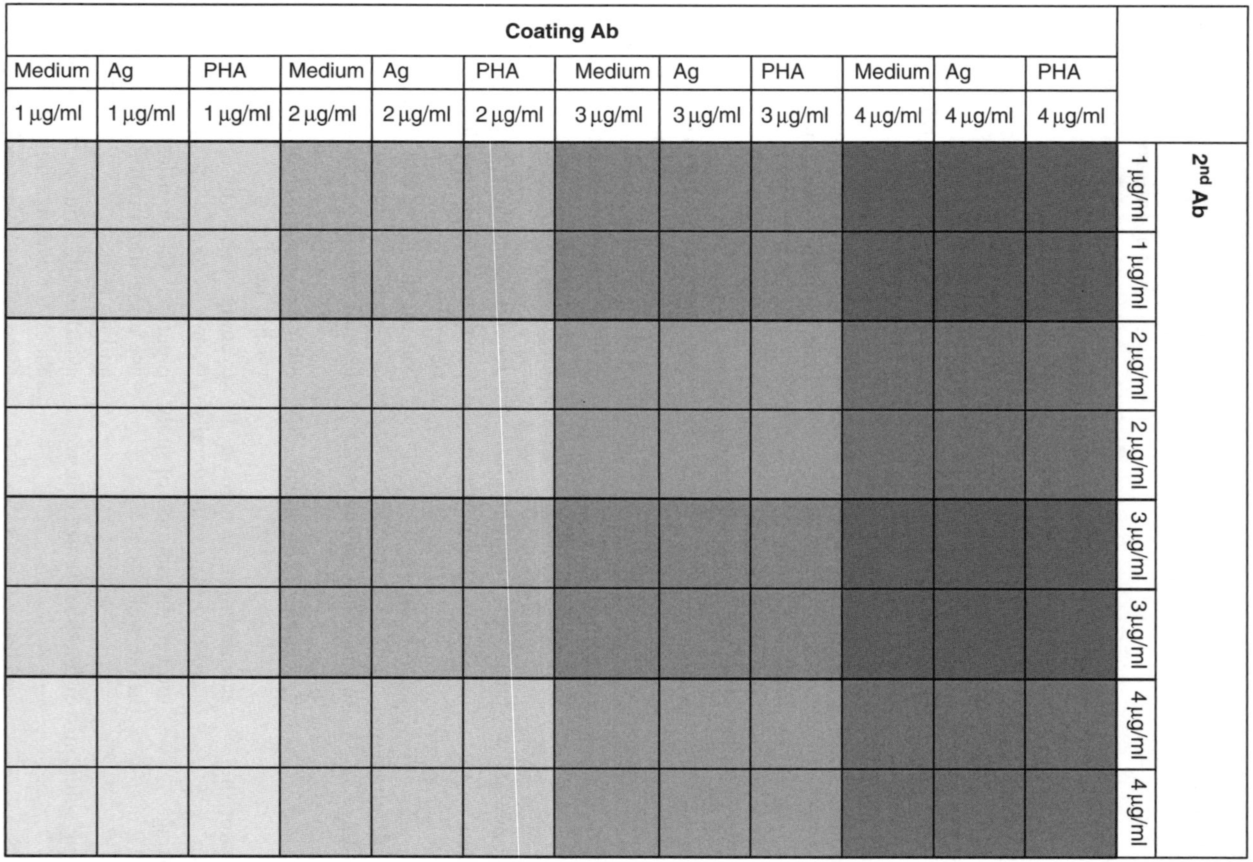

Figure 9.3 Plate layout for an example Ab cross-titration. Horizontally the coating antibody is titrated and vertically the second antibody

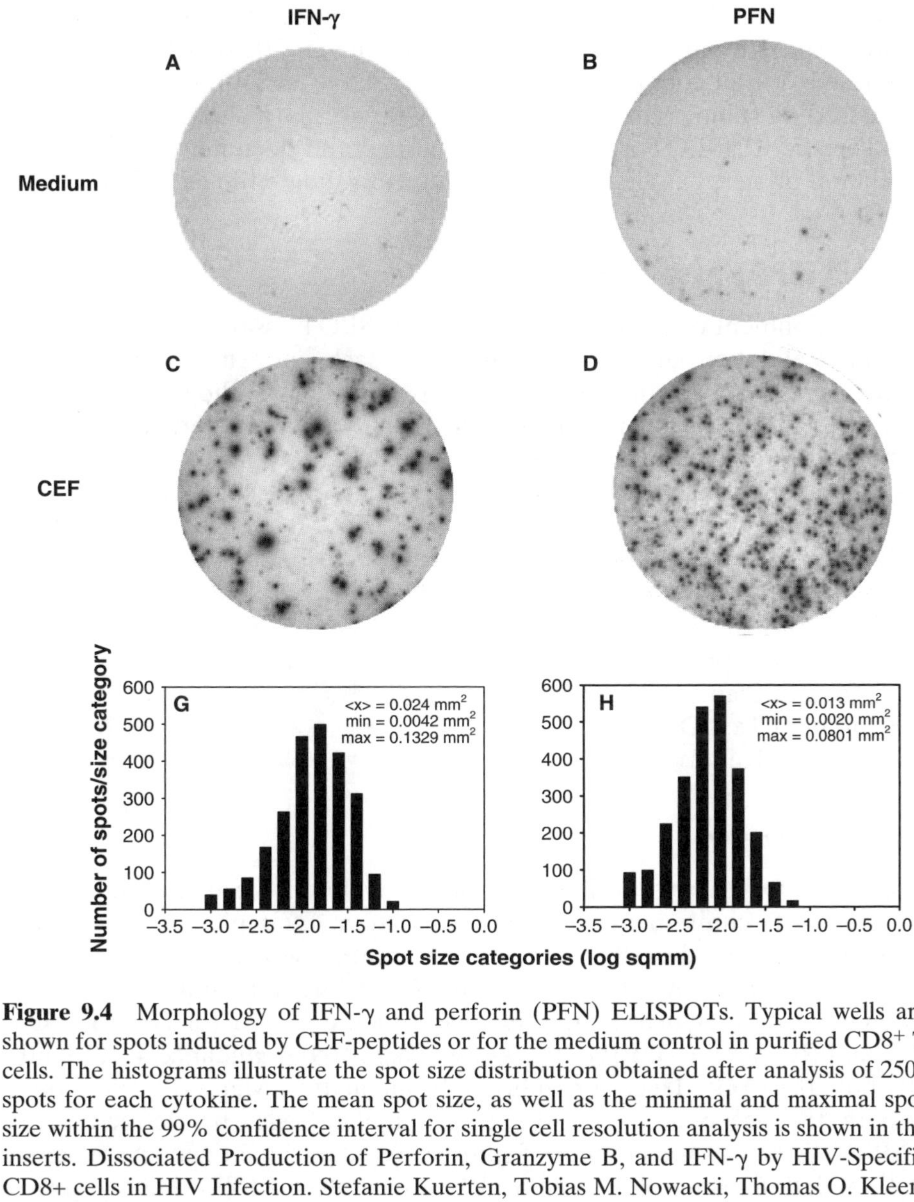

Figure 9.4 Morphology of IFN-γ and perforin (PFN) ELISPOTs. Typical wells are shown for spots induced by CEF-peptides or for the medium control in purified $CD8^+$ T cells. The histograms illustrate the spot size distribution obtained after analysis of 2500 spots for each cytokine. The mean spot size, as well as the minimal and maximal spot size within the 99% confidence interval for single cell resolution analysis is shown in the inserts. Dissociated Production of Perforin, Granzyme B, and IFN-γ by HIV-Specific CD8+ cells in HIV Infection. Stefanie Kuerten, Tobias M. Nowacki, Thomas O. Kleen, Robert J. Asaad, Paul V. Lehmann, Magdalena Tary-Lehmann, AIDS Research and Human Retroviruses, 2008, 24(1): 62–71.

cytokines and this will also influence the size of the spots measured in the ELISPOT (Guerkov et al, 2003). The size of the spots measured in the ELISPOT is also influenced by the product measured (Figure 9.5) even if it is produced by the same cell type. The size and the quality of the spot will be important when the ELISPOT assay will be counted. The counting should be performed by an automated system (Lehmann 2005) that is regularly calibrated. Visual counting of spots leads to large variations and inaccuracies based on subjectivity. The number of variables which impact the result of a specific ELISPOT assay must be defined and documented before the validation process can be initiated in order to validate the ELISPOT assay.

9.5 ELISPOT assay validation

The development of the antigen specific ELISPOT assay identifies the test system. Validation should be planned for each antigen and cytokine to be measured or any other measured products. Assay validation is divided into two phase. Phase I includes the Qualification Experiments, which establish the precise parameters and acceptance criteria for the validation experiment. Phase II then includes the Validation Experiment and aims to qualify one or more reference samples for use in a regulated study. By including both of these phases, the validation covers, but is not limited to, the following aspects of the assay structure

- Assay Development
- Validation Plan
- Validation Test Data
- Validation Report

Testing parameters

The ELISPOT Assay Validation using a Reference Sample will follow the standard requirements for assay validation. Examples of such guidelines include those described by the FDA in http://www.fda.gov/cder/ guidance/ichq2a.pdf (July of 2006) titled "Guideline for Industry Text on Validation of Analytical Procedures" and by the Center for Drug Evaluation and Research (CDER) http://www.fda.gov/ cder/guidance/4252fnl.htm (July of 2006) "Guidance for Industry Bio-analytical Method Validation".

With the acceptance that the term "analyte" can be defined in an ELISPOT assay as the number of spots secreted by an antigen (Ag) specific cell, validation testing should be broken down into the following categories:

Accuracy

"The accuracy of an analytical procedure expresses the closeness of agreement between the value which is accepted either as a conventional true value or an accepted reference value and the value found." (Note: definitions of test parameters in parentheses are citations from the FDA guideline Q2A).

This could be defined with two different assay systems, either ICS or ELISPOT. The different sensitivity of the two assay systems makes this challenging, but can be performed if the frequency of the responding population is high enough. These types of measurements have been performed and reported in the literature, showing that the ELISPOT assay can, in fact, detect human IFN-γ produced by individual responder T cells, and therefore can accurately measure frequencies of IFN-γ-producing cells (Heeger et al, 1999; Helms et al, 2000). In mice, pigs and rhesus macaques similar studies were performed with similar findings (Karulin et al, 2000; Moretto et al, 2000; Pahar et al, 2003; Rau et al, 2006).

Precision

"The precision of an analytical procedure expresses the closeness of agreement (degree of scatter) between a series of measurements obtained from multiple sampling of the same homogeneous sample under the prescribed conditions." (Note: definitions of test parameters in parentheses are citations from the FDA guideline Q2A). The following precision parameters can be tested: repeatability as well called intra-assay precision (for example well to well and day to day), intermediate precision (for example plate to plate, different analysts, and different equipment) and reproducibility (laboratory to laboratory). Precision may be established without the availability of a "gold" standard as it represents the scatter of the data rather than the exactness (accuracy) of the reported result. The ELISPOT system's capability to use frozen cells makes it possible to generate reference samples. The reference sample could be derived by leucopheresis. Up to 20 billion PBMC from a single human donor at a single time point can be extracted in this way. Aliquoting and freezing such large leucopheresis samples (typically at 10 million PBMC per vial) therefore could provide up to 2000 reference vials/samples for 2000 assays! Therefore sufficient cells for the entire duration of a trial are created. Precision of a specific assay could then be determined by running qualification experiments with the reference sample and derive out of these experiments the precision values. This value can then be used for the validation experiment. For example for a qualification experiment in total three plates could be tested performed by three different technicians run on three different days.

Specificity

"Specificity is the ability to assess unequivocally the analyte in the presence of components which may be expected to be present. Typically these might include impurities, degradants, matrix, etc." (Note: definitions of test parameters in parentheses are citations from the FDA guideline Q2A). The specificity of an assay is the capability of the assay to differentiate similar analyte or interference from matrix elements that could have a positive of negative effects on the assay value. The cell population tested in ELISPOT is specific to a certain Ag. ELISPOT determines if an individual subject has T cells with specificity that recognizes the test Ag. The secretion of the cytokine of interest is induced and captured in the ELISPOT plate only if the T cells are specific to the tested Ag. T cells which do not get activated by the test Ag will not secrete the tested cytokine. Specificity is determined by the tested cell population which results in spot production above medium (Y spot number) after exposure to a given Ag (X number of spots). If the tested population is not recognizing the test Ag it will not secrete the cytokine in spot numbers exceeding that of the above medium.

Specificity of the response can therefore be represented as $[X-(Y+O+2SD \text{ of } Y)]$ Whererby

X: Average spot number induced by a given Ag.
Y: Average spot number in the medium control.
O: Average spot number in the negative control no cells added.

Specificity of this assay will then be determined by running qualification experiments with the reference sample to be validated. The specificity value will be derived from these experiments. This value will then be used for the validation experiment.

Linearity

"Linearity is the ability of the assay to return values that are directly proportional to the concentration of the analyte in the sample". (Note: definitions of test parameters in parentheses are citations from the FDA guideline Q2A). In ELISPOT, this refers to the titrated test population's (T cells') response (IFN-γ spots) to the same Ag. Linearity can be determined, but it is not useful for ELISPOT assays in a clinical setting. The experiment requires too much blood from each subject sample. T cells and antigen presenting cells (APCs) would need to be separated by negative selection from the same donor. Purified T cells could be mixed in serial dilution with APC, keeping the total number of cells constant, i.e., 300 000 cells/well. This experiment is very labor intensive, expensive and requires considerable numbers of reference sample cells. This has been explored in multiple scientific publications proving the single cell resolution of the ELISPOT test method (Heeger et al, 1999; Helms et al, 2000; Hesse et al, 2001; Kreher et al, 2003).

Limits of Detection

The Limits of Detection (LOD) are the highest and the lowest number of spots which can be detected and counted in a given well. LOD is the simple detection of spots in a given well. This detection is not test antigen specific. Lowest Limits of Detection (LLOD) are the lowest amount of spots, which can be detected. The detection limit is a low number of spots that is statistically distinguishable from background or negative control (O) where no cells are plated in a given well (LLOD $= O + 2$SD of O).

Highest Limit of Detection (HLOD) is the number of spots which can be counted with good confidence after mitogen stimulation in that specific assay system, not reaching TNTC (too numerous to count), by either an automated or manual method of image analysis (HLOD $= H$).

Whereby

H: Highest spot number countable.

Limits of Quantitation (Ag specific response)

'The quantitation limit of an individual analytical procedure is the lowest amount of analyte in a sample which can be quantitatively determined with suitable precision and accuracy." (Note: definitions of test parameters in parentheses are citations from the FDA guideline Q2A).

Limit of quantitation (LOQ) is the lowest and highest number of spots by a subject (per donor, per individual) that can be quantitatively determined with suitable precision and accuracy to the test Ag. The lower limit of quantitation is often defined by an arbitrary cut-off such as a ratio of signal-to-noise, equal to X, or a greater value then the mean of the medium control plus 2 times the standard deviation of the medium control values. Highest Limit of Quantification is the number of spots which can be counted with good confidence after antigen stimulation, not reaching TNTC (spot density is too numerous to count), by an image analyzer. If TNTC is reached, the response is positive but no value can be given to this Ag specific response.

Lowest Limit of Quantitation (LLOQ) $= X - (Y + O + 2\text{SD of } Y)$

Highest Limit of Quantitation (HLOQ) $= H$

Range

"The range of an analytical procedure is the interval between the upper and lower concentration (amounts) of analyte in the sample (including these concentrations) for which it has been demonstrated that the analytical procedure has a suitable level of precision, accuracy and linearity." (Note: definitions of test parameters in parentheses are citations from the FDA guideline Q2A). Range is the concentrations of analyte or assay values between the low and high limits of quantitation.

The range in an ELISPOT assay can therefore be defined by the LLOQ value and the HLOQ value. See above.

Test procedures

Before any and all test procedures start, a validation protocol needs to be written. It will contain test procedure, plate layout, tables and all necessary information to successfully run the experiments. This plan needs to be approved and signed. Only after this is done can the test procedures be performed.

Phase I – qualification experiments

An example of the parameters requiring definition in a qualification experiment set are presented here:

- Reference sample to be tested from subject: XX
- Bleed date that sample was obtained: YY
- number of cells to be plated per well: 300 000
- Antigen to be used and concentration: AA at 10 μg/mL.
- Assay will be performed according to standard operating procedure number (worked out during developmental experiments). This will include number or replicates on the plate, number of technicians repeating the work, number of days on which the repeats are done and so forth.

All of the data to be generated need to be added into predefined tables from the validation plan. The prepared tables should contain the raw spot number per well, the mean or median and the %RSD or other chosen statistical read out. The below list of tables could be prepared:

- For the data set – Repeatability (well to well)
- For the data set – Intermediate precision (plate to plate)
- For the data set – Intermediate precision (for example technician to technician)
- For the data set – Reproducibility (day to day)

These qualification experiments will then define the acceptance criteria for the reference sample to the given antigen.

Phase II – validation experiment

The qualification experiments' acceptance criteria will be known before the validation experiment is started and documented in the predefined table before the validation experiments is performed. In the validation experiments, the values for the validation results will be recorded. If they fulfill the predefined acceptance criteria, the validation of the assay with the given reference sample is successful, the validation report can be approved, and the reference samples can be used for comparison in the pre-clinical/clinical study.

Statistics

The *relative standard deviation* (RSD) is useful for comparing the uncertainty between different measurements of varying absolute magnitude. The RSD is calculated from the standard deviation, s, and the mean, x (of all measurements of spot numbers induced by a given antigen per well form the same sample) and is commonly expressed as a percentage (%).

The %RSD can be used when discussing the normal distribution for positive mean values with the standard deviation significantly less than the mean. However when the mean value is near zero, the %RSD is sensitive to change in the standard deviation and limits its usefulness. This captures the finding that the lower the number of spots per well, the higher the %RSD (Figure 9.5).

Medium responses (the spontaneous release of measured product by the plated cells) should not be expressed by %RSD, but the value should be

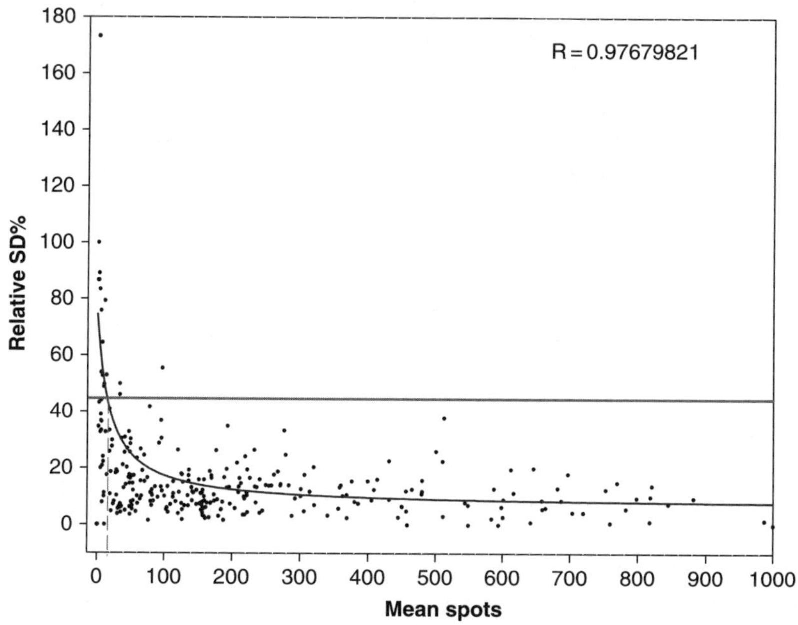

Figure 9.5 Statistical analysis of recall antigen-induced IFN-γ. The intra-assay precision/repeatability is compared. The %RSD was calculated and plotted against the corresponding mean spot numbers

expressed in arithmetic mean or median numbers of spots with a standard deviation value (Example: 15 spots ±7 spots). For example if triplicates are used for given antigens in the human IFN-γ ELISPOT assay as shown in Figure 9.5 the minimal number of Ag-induced spots needs to be higher than 20 spots/well. The statistical analysis of recall antigen induced human IFN-γ and IL-2 ELISPOT assays in fresh and frozen samples has been previously determined (Kreher et al, 2003) Other statistical analyses of the ELISPOT assay are also available (Cox et al, 2005; Gebauer et al, 2002; Janetzki et al, 2005; Maecker et al, 2005; Smith et al, 2001).

Maintaining a validated state

Since there is a great amount of effort put forth to standardize and validate reference samples for regulated ELISPOT assays, it is equally important to maintain this validated state. When a scientist defines his or her testing parameters, i.e., plate type, incubation periods, instrumentation, etc., these defined parameters must remain the same throughout the life of the validation and testing of the pre-clinical/clinical samples. If parameters need to be changed, then dependent on the extent of the change, partial or full re-validation may be required. For example, if plate type is changed, then testing to determine the impact of this plate change must be performed and documented. If different reference samples are used, then a full re-validation is required. If different antigens or cytokines are used even with the same reference samples, then a full new validation must be performed. Depending on the importance of the regulated study, i.e., pilot versus pivotal study, the governing regulatory agency may require further validation.

Special thanks

We would like to give special thanks to the following: Shauna Heeger, Director, Business Development, CTL, CTL Clients, and CTL Academic Colleagues.

References

Anthony DD and Lehmann PV (2003). T-cell epitope mapping using the ELISPOT approach. *Methods*, **29**, 260–269.

Anthony DD, Valdez H, Post AB, Carlson NL, Heeger PS and Lehmann PV (2002). Comprehensive determinant mapping of the hepatitis C-specific CD8 cell repertoire reveals unpredicted immune hierarchy. *Clin Immunol*, **103**, 264–276.

Carter LL and Swain SL (1997). Single cell analyses of cytokine production. *Curr Opin Immunol*, **9**, 177–182.

Cox JH, Ferrari G, Kalams SA, Lopaczynski W, Oden N and D'souza MP (2005). Results of an ELISPOT proficiency panel conducted in 11 laboratories participating in international human immunodeficiency virus type 1 vaccine trials. *AIDS Res Hum Retroviruses*, **21**, 68–81.

Currier JR, Kuta EG, Turk E, Earhart LB, Loomis-Price L, Janetzki S, Ferrari G, Birx DL, and Cox JH (2002). A panel of MHC class I restricted viral peptides for use as a quality control for vaccine trial ELISPOT assays. *J Immunol Methods*, **260**, 157–172.

Czerkinsky C, Andersson G, Ekre HP, Nilsson LA, Klareskog L and Ouchterlony O (1988). Reverse ELISPOT assay for clonal analysis of cytokine production. I. Enumeration of gamma-interferon-secreting cells, *J Immunol Methods*, **110**, 29–36.

Czerkinsky CC, Nilsson LA, Nygren H, Ouchterlony O and Tarkowski A (1983). A solid-phase enzyme-linked immunospot (ELISPOT) assay for enumeration of specific antibody-secreting cells, *J Immunol Methods*, **65**, 109–121.

Forsthuber T, Yip HC and Lehmann PV (1996). Induction of TH1 and TH2 immunity in neonatal mice, *Science*, **271**, 1728–1730.

Gebauer BS, Hricik DE, Atallah A, Bryan K, Riley J, Tary-Lehmann M, Greenspan NS, Dejelo C, Boehm BO, Hering BJ and Heeger PS (2002). Evolution of the enzyme-linked immunosorbent spot assay for post-transplant alloreactivity as a potentially useful immune monitoring tool, *Am J Transplant*, **2**, 857–866.

Guerkov RE, Targoni OS, Kreher CR, Boehm BO, Herrera MT, Tary-Lehmann M, Lehmann PV and Schwander SK (2003). Detection of low-frequency antigen-specific IL-10-producing CD4(+) T cells via ELISPOT in PBMC: cognate vs. nonspecific production of the cytokine, *J Immunol Methods*, **279**, 111–121.

Heeger PS, Greenspan NS, Kuhlenschmidt S, Dejelo C, Hricik DE, Schulak JA, and Tary-Lehmann M (1999). Pretransplant frequency of donor-specific, IFN-gamma-producing lymphocytes is a manifestation of immunologic memory and correlates with the risk of posttransplant rejection episodes, *J Immunol*, **163**, 2267–2275.

Helms T, Boehm BO, Asaad RJ, Trezza RP, Lehmann PV and Tary-Lehmann M (2000). Direct visualization of cytokine-producing recall antigen-specific CD4 memory T cells in healthy individuals and HIV patients, *J Immunol*, vol. 164, no. 7, pp. 3723–3732.

Hesse MD, Karulin AY, Boehm BO, Lehmann PV and Tary-Lehmann M (2001). A T cell clone's avidity is a function of its activation state, *J Immunol*, **167**, 1353–1361.

Janetzki S, Cox JH, Oden N and Ferrari G (2005). Standardization and validation issues of the ELISPOT assay, *Methods Mol Biol*, **302**, 51–86.

Karulin AY, Hesse MD, Tary-Lehmann M and Lehmann PV (2000). Single-cytokine-producing CD4 memory cells predominate in type 1 and type 2 immunity, *J Immunol*, **164**, 1862–1872.

Kleen TO, Asaad R, Landry SJ, Boehm BO and Tary-Lehmann M (2004). Tc1 effector diversity shows dissociated expression of granzyme B and interferon-gamma in HIV infection, *AIDS*, **18**, 383–392.

Kleen TO, Kadereit S, Fanning LR, Jaroscak J, Fu P, Meyerson HJ, Kulchycki L, Slivka LF, Kozik M, Tary-Lehmann M and Laughlin MJ (2005). Recipient-specific tolerance after HLA-mismatched umbilical cord blood stem cell transplantation, *Transplantation*, **80**, 1316–1322.

Kreher CR, Dittrich MT, Guerkov R, Boehm BO and Tary-Lehmann M (2003). CD4+ and CD8+ cells in cryopreserved human PBMC maintain full functionality in cytokine ELISPOT assays, *J Immunol Methods*, **278**, 79–93.

Lehmann PV (2005). Image analysis and data management of ELISPOT assay results, *Methods Mol Biol*, **302**, 117–132.

Maecker HT, Moon J, Bhatia S, Ghanekar SA, Maino VC, Payne JK, Kuus-Reichel K, Chang JC, Summers A, Clay TM, Morse MA, Lyerly HK, DeLaRosa C, Ankerst

DP and Disis ML (2005). Impact of cryopreservation on tetramer, cytokine flow cytometry, and ELISPOT, *BMC Immunol*, **6**, 17.

McCutcheon M, Wehner N, Wensky A, Kushner M, Doan S, Hsiao L, Calabresi P, Ha T, Tran TV, Tate KM, Winkelhake J and Spack EG (1997). A sensitive ELISPOT assay to detect low-frequency human T lymphocytes, *J Immunol Methods*, **210**, 149–166.

McHugh S, Deighton J, Rifkin I and Ewan P (1996). Kinetics and functional implications of Th1 and Th2 cytokine production following activation of peripheral blood mononuclear cells in primary culture, *Eur J Immunol*, **26**, 1260–1265.

Moretto WJ, Drohan LA and Nixon DF (2000). Rapid quantification of SIV-specific CD8 T cell responses with recombinant vaccinia virus ELISPOT or cytokine flow cytometry, *AIDS*, **14**, 2625–2627.

Ott PA, Herzog BA, Quast S, Hofstetter HH, Boehm BO, Tary-Lehmann M, Durinovic-Bello I, Berner BR and Lehmann PV (2005). Islet-cell antigen-reactive T cells show different expansion rates and Th1/Th2 differentiation in type 1 diabetic patients and healthy controls, *Clin Immunol*, **115**, 102–114.

Pahar B, Li J, Rourke T, Miller CJ and McChesney MB (2003). Detection of antigen-specific T cell interferon gamma expression by ELISPOT and cytokine flow cytometry assays in rhesus macaques, *J Immunol Methods*, **282**, 103–115.

Rau H, Revets H, Balmelli C, McCullough KC, and Summerfield A (2006). Immunological properties of recombinant classical swine fever virus NS3 protein in vitro and in vivo, *Vet Res*, **37**, 155–168.

Rininsland FH, Helms T, Asaad RJ, Boehm BO and Tary-Lehmann M (2000). Granzyme B ELISPOT assay for ex vivo measurements of T cell immunity, *J Immunol Methods*, **240**, 143–155.

Schielen P, van Rodijnen W, Tekstra J, Albers R and Seinen W (1995). Quantification of natural antibody producing B cells in rats by an improved ELISPOT technique using the polyvinylidene difluoride membrane as the solid support, *J Immunol Methods*, **188**, 33–41.

Sedgwick JD and Holt PG (1983). A solid-phase immunoenzymatic technique for the enumeration of specific antibody-secreting cells, *J Immunol Methods*, **57**, 301–309.

Shaw RD, Merchant AA, Groene WS and Cheng EH (1993). Persistence of intestinal antibody response to heterologous rotavirus infection in a murine model beyond 1 year, *J Clin Microbiol*, **31**, 188–191.

Smith JG, Liu X, Kaufhold RM, Clair J and Caulfield MJ (2001). Development and validation of a gamma interferon ELISPOT assay for quantitation of cellular immune responses to varicella-zoster virus, *Clin Diagn Lab Immunol*, **8**, 871–879.

Tanguay S and Killion JJ (1994). Direct comparison of ELISPOT and ELISA-based assays for detection of individual cytokine-secreting cells, *Lymphokine Cytokine Res*, **13**, 259–263.

10 IFN-γ ELISPOT assay validation

Manjula Reddy[1], Jackson Wong, Charles Pendley and Uma Prabhakar

Centocor Research and Development, Inc., 145 King of Prussia Road, Radnor, PA 19087, USA

10.1 Introduction

Cytokines are critical regulators of pro-inflammatory and anti-inflammatory immune responses. T helper 1 (T_H1) cytokines, such as interferon-gamma (IFN-γ), interleukin (IL)-12, and IL-2 promote cell-mediated responses, while T helper 2 (T_H2) cytokines, such as, IL-4, IL-5, IL-6, and IL-10 are anti-inflammatory and favor humoral responses (Bennouna et al, 2002; Navikas and Link, 1996; Okamoto et al, 2004). A predominant T_H1 response is associated with autoimmune diseases whereas a predominant T_H2 response is associated with allergic responses and systemic autoimmunity (Mehlis and Gordon, 2004). As potential new immunomodulating therapies designed to modify T_H1-type and T_H2-type responses enter clinical trials, validated assays for use in clinical sample testing are needed to monitor the response of patients to treatment. The enzyme-linked

[1] To whom correspondence should be addressed.

Validation of Cell-Based Assays in the GLP Setting: A Practical Guide Editors, Uma Prabhakar and Marian Kelley

immunospot or ELISPOT assay can be used to measure cellular immunogenicity, including the T_H1 and T_H2 responses. In the ELISPOT assay, secreted cytokines immediately adhere to the solid phase of the ELISPOT assay plate and are not consumed by other cells in the culture or affected by simultaneously produced antagonists. Thus, the ELISPOT assay measures the net production of cytokines more stringently than other methods, such as enzyme-linked immunosorbent assays. It also allows evaluation of cryopreserved human lymphocytes without loss of functional responses with greater sensitivity and precision facilitated by automated imaging systems for spot analysis. All these properties of the ELISPOT assay make it useful for monitoring treatment effects in clinical trials (Bennouna et al, 2002; Kreher et al, 2003, Farina et al, 2003).

The utility of the IFN-γ ELISPOT assay in detecting the frequency of antigen-specific T lymphocytes in autoimmune disease models and immune responses in patients with infectious, autoimmune, and neoplastic diseases undergoing immunotherapeutic treatments has been described in several studies (Currier et al, 2002; Hedman et al, 2006; Jansson et al, 2003; Letsch and Scheibenbogen, 2003; Pelfrey et al, 2000; Ronnelid et al, 1998; Smith et al, 2001). It was reported that less than 0.05% of peripheral blood mononuclear cells (PBMCs) spontaneously released IFN-γ. Further, IFN-γ was shown to be secreted by 26.9% of CD4+, 48.7% of CD8+, and 15.9% of CD14+ and 8.4% of other cells (Hagiwara et al, 1995). IFN-γ plays key roles in host defense by exerting anti-viral, anti-proliferative and immunoregulatory activities (Boehm et al, 1997). The production of IFN-γ was reported to be upregulated synergistically by IL-12 and IL-18 (Tominaga et al, 2000) and therefore it is relevant to measure modulation of IFN-γ responses in clinical studies involving anti-IL-12 therapy.

Polyclonal activators such as: (1) phytohemagglutinin-P (PHA-P), (2) phorbol 12-myristate 13-acetate (PMA) + ionomycin, and (3) the superantigens, staphylococcal enterotoxins A and B (SEA and SEB) were used to validate the IFN-γ ELISPOT assay. Lectins such as PHA, bind to specific glycoproteins on the cell surface and mimic many of the same pathways initiated by T cell receptor engagement of the Ag-MHC (where MHC means major histocompatibility complex) complex. PHA-P, a T cell mitogen has been shown to cause a significant increase in the number of cells secreting IL-2, IL-4, and IFN-γ (Hagiwara et al, 1995) and is therefore an ideal control for measuring both T_H1 and T_H2 cell responses. The phorbol esters such as PMA along with the calcium ionophore ionomycin, broadly activate lymphocytes and are useful to measure cytokine secretion from various T cell subpopulations (Maino, 1998; Pala et al, 2000). The superantigens bind directly to MHC Class II molecules on antigen-presenting cells and this complex binds to different Vβ regions of the T cell antigen receptor causing T cell activation (Johnson et al, 1991; Kotzin et al, 1993).

Staphylococcal superantigens stimulate subsets of T cells expressing specific Vβ receptors. Between 5 to 30% of the T cell repertoire may be stimulated by a particular superantigen, depending on the relative frequency of the Vβ targeted (Kotzin et al, 1993; Marrack and Kappler, 1990). PBMCs stimulated by superantigens produce both T_H1 and T_H2 cytokines and can be used for the longitudinal monitoring of immune responses in clinical study samples. Evaluating reactivity to superantigens is particularly relevant in inflammatory skin diseases such as atopic dermatitis and psoriasis that are reported to be associated with *Staphylococcus aureus* (Zollner et al, 2000). The validation of an IFN-γ ELISPOT assay for use in clinical sample testing is described in this chapter including appropriate methods of PBMC cryopreservation and thawing. Various validation parameters that are relevant for implementation of the assay are described as well. These parameters are discussed in more detail in the preceding chapter.

10.2 Methods

Sample collection and isolation of PBMCs

Whole blood was collected from normal volunteers into 10 mL Vacutainer tubes containing the anticoagulant, potassium EDTA. PBMCs were isolated on the day of sample collection or after overnight shipment using Ficoll-Plus density gradient centrifugation (Amersham Biosciences, Piscataway, NJ). Any contaminating erythrocytes in the PBMC suspension were removed by lysis using the red blood cell (RBC) lysing kit from R&D Systems. RBC aggregation and contamination of the PBMCs is more common in stored or shipped blood and increases with storage time. Briefly, PBMCs were washed with 1X wash buffer from the RBC lysing kit, resuspended in appropriate amount of culture medium, and cell viability determined using the Guava PCA system (Guava Technologies, MN). PBMCs were used in the ELISPOT validation with or without cryopreservation in de-complemented human AB serum containing 10% DMSO. Since functional preservation of PBMCs is vital for a successful cytokine response, proper cryopreservation and thawing methods are important and are described below.

Cryopreservation of PBMCs to be used in the ELISPOT assay

Careful cryopreservation of the PBMCs is important for functional preservation of the cells and a successful cytokine response. To cryopreserve the cells, PBMCs suspended in decomplemented human AB serum at room temperature were mixed with an equal volume of 20% DMSO in human AB serum pre-warmed to room temperature by drop-wise addition to avoid osmotic shock and loss of viability and function of the cells.

Aliquots of the PBMC suspension in cryovials were placed in a Nalgene Cryo Freezing container filled with *n*-propanol at room temperature. The freezing container was placed in –80 °C overnight. The frozen cells were then transferred to a liquid nitrogen freezer for long-term storage.

Thawing of cryopreserved PBMCs for ELISPOT assay

Careful thawing of the PBMCs is also important for functional preservation of the cells and a successful cytokine response. Cryopreserved PBMCs were rapidly thawed in a 37 °C water bath and cells were transferred to culture medium containing 10% heat-inactivated fetal bovine serum (FBS) (JRH Biosciences, Lenexa, KS from a pre-tested lot) and pencillin-streptomycin-L-glutamine (Invitrogen, Grand Island, NY). The cells were washed and resuspended in fresh medium and the viable cell count was determined.

IFN-γ ELISPOT protocol

The outline of the general ELISPOT procotol includes: (1) coating 96-well microtiter plates with purified cytokine-specific antibody; (2) blocking the plate to prevent nonspecific binding of random proteins, (3) incubating cells with and without antigens, (4) lysing cells with detergent, (5) adding a labeled second antibody, and (6) detecting the antibody–cytokine complex using an enzyme–substrate reaction resulting in a colored spot image that can be quantified microscopically or electronically. Each spot represents one single cell secreting the cytokine of interest. The antigen-specific precursor frequency is determined by dividing the number of spots by the number of cells plated in the wells (Clay et al, 2001; Letsch and Scheibenbogen, 2003).

Using sterile technique, MultiScreen™ – IP plates (Millipore Corporation, Billerica, MA) are coated overnight in the refrigerator with anti-IFN-γ capture antibody diluted into calcium and magnesium-free Dulbecco's phosphate-buffered saline (Cellgro, Herndon, VA) according to the manufacturer's instructions. The plates are blocked by incubation with culture medium and then, washed. Aliquots of the antigen solution and the cell suspension are added to the cells in triplicate. The optimal number of PBMCs for the IFN-γ ELISPOT assay varies from 10 000 to 25 000 per well for cryopreserved PBMCs stimulated with PMA + ionomycin to 100 000 to 200 000 per well for stimulation with PHA-P or the superantigens depending on the expected responses. When low frequency responses are anticipated, 300 000 cells per well should be used for PHA-P or superantigen stimulation and 25 000 for PMA + ionomycin.

The ELISPOT cell culture plates are incubated for 20 hours at 37°C in an atmosphere of 5% CO_2 and 99% humidity. Then, the plates are washed twice with distilled water and three times with 1X Dulbecco's phosphate-buffered saline (DPBS) with 0.05% Tween 20 and biotinylated detection antibody in dilution buffer (calcium and magnesium-free DPBS with 10% FBS) is added. After incubation for 2 hours at room temperature, the plates are washed and avidin horse radish peroxidase (HRP) (BD Pharmingen, San Diego, CA) is added. The plates are incubated for an additional hour at room temperature, washed, and the substrate (AEC substrate, BD Biosciences Pharmingen, San Diego, CA) added. Color development is monitored for the appearance of bright brown spots in the positive control wells vs. the negative control wells (wells with stimulation) and is allowed to proceed for no longer than 5 minutes. The reaction is terminated by washing the plates with distilled water. The plates are air dried in the dark for 2 hours or overnight at room temperature in the biological hood, scanned using the ImmunoSpot analyzer (BD Biosciences, San Jose, CA), and counted using ImmunoSpot version 3.1 software (Cellular Technology, Ltd., Cleveland, OH). Plates can be stored for a few months wrapped in aluminum foil.

ELISPOT result output and data management tools

The auto-count feature of the ImmunoSpot software counts spots in all wells or selected wells of the 96-well plate after counting parameters have been established. The results, number of spots and spot size, can be imported into spreadsheet software such as Microsoft Excel for calculating the mean of the replicates, the standard deviation (SD) and percent coefficient of variability (%CV) as well as other parameters associated with the experiment such as titration of antigens, cells etc.

Two other software modules, SpotMap™ and ImmunoSpot quality control (QC) (Cellular Technology Ltd., Cleveland, OH) can be used when clinical samples are being assayed. SpotMap™ is a data management tool for high volume ELISPOT data processing that allows the user to link the spot counts with donor, antigen, and reagent information and to process/manage all these data. The QC module can be used to correct counts from wells with large clusters of spots or indiscrete spots that can occur when cryopreserved PBMCs are used.

Data analysis and result interpretation

The number of spots and spot size are the output generated in an ELISPOT assay. Each spot in the ELISPOT assay typically represents cytokine secreted from a single PBMC and the number of spots represents the frequency of cytokine-producing PBMCs in each well of the plate. The spot counts can be represented as the number of spots per total number of cells added to the well or can be used to calculate the number of spot forming cells (SFC) per million PBMCs. The size of the spots reflects the amount of cytokine produced during the assay because released cytokines are continuously captured on the membrane surrounding the secreting cell, which permits quantification of cytokine output per cell (Helms et al, 2000) and reflects the kinetics of cytokine production. The ImmunoSpot software uses mean spot size to estimate cytokine secretion for the entire cell population in the assay well. The spot size measurement is not described in this validation because it is more relevant when specific antigens are used or when the amount of secreted product needs to be quantified.

Results are presented as the number of spots per total number of cells added to the wells. When samples from clinical studies are tested, the final results are represented as spot forming cells (SFC) per million PBMCs.

10.3 Validation of the IFN-γ ELISPOT assay

Validation of the IFN-γ ELISPOT assay (measuring IFN-γ secretion by stimulated PBMCs) focused on the following parameters: specificity of anti-cytokine antibody pairs, cell density, positive control antigen concentration, stability of the profile of cytokine responses with different lots of antigens using fresh and cryopreserved PBMCs, and the effects of shipment and various storage conditions on the ability of the PBMCs to secrete IFN-γ in response to stimulation. The stability of the IFN-γ cytokine response in PBMCs from normal donors after long term freezing was also addressed. Assay precision was evaluated using overnight shipped and cryopreserved PBMCs, by determining the intra-assay (variability in replicates), inter-assay (plate-to-plate variability and within day variability), and inter-operator variability. Data and sample acceptance criteria were addressed by evaluating the cytokine responses of PBMCs isolated and cryopreserved from shipped whole blood samples from normal donors. All validation parameters were studied using stimulated PBMCs because unstimulated cells spontaneously produce low or undetectable levels of IFN-γ. The validation of the IFN-γ ELISPOT assay using the T cell activator PHA-P as an example, is described in this chapter.

Validation of the positive control antigens for optimal IFN-γ secretion

Part 1: studies with fresh cells

Reagent validation and specificity

The specificity of the ELISPOT assay for IFN-γ was examined by switching the detection antibodies in the cytokine antibody pairs. Freshly isolated PBMCs from a normal donor were plated into wells coated with the anti-IFN-γ capture antibody and cultured with and without PHA. The results

Table 10.1 Specificity of the IFN-γ capture and detection antibody pair[a,b]

	Capture antibody	Detection antibody	Mean	SD	%CV
PHA-P	IFN-γ	IFN-γ	185	24.01	12.95
PMA + ionomycin	IFN-γ	IFN-γ	236	33.78	14.31
PHA-P	IFN-γ	IL-5	0	0.58	
PMA + ionomycin	IFN-γ	IL-5	0	0.58	
PMA + ionomycin	IFN-γ	None, Streptavidin-HRP and substrate only	0	0.00	

[a] Values are the number of IFN-γ spots per 100 000 PBMCs.
[b] PHA-P indicates phytohemagglutinin-P; IFN-γ, interferon-γ, IL-5, interleukin-5, HRP, horse radish peroxidase.

in Table 10.1 indicate that the IFN-γ ELISPOT assay is specific because no ELISPOTs developed when unmatched pairs of anticytokine antibodies were used for capture and detection and there is no cross-reactivity with the anti-IL-5 detection antibody.

Optimal cell density for measuring IFN-γ with PHA-P stimulation

The optimal cell density for PHA-P induced IFN-γ response was determined by plating PBMCs isolated from a normal donor peripheral blood in triplicate immediately after blood collection at cell numbers ranging from 250 to 100 000 cells. The cells were stimulated with 5 μg/mL PHA-P. Though highly variable IFN-γ responses were observed at 50 000 cells, a stronger response was observed with 100 000 cells (441 ± 22.5 IFN-γ spots and %CV of 5.10 between replicates as shown in Figure 10.1).

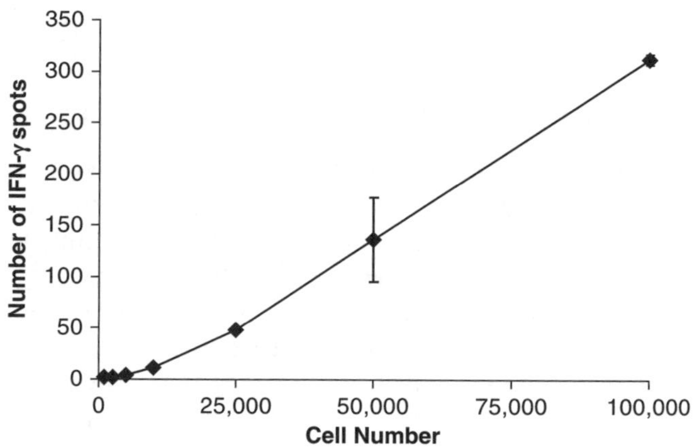

Figure 10.1 Secretion of IFN-γ in response to PHA stimulation for PBMCs from two normal donors

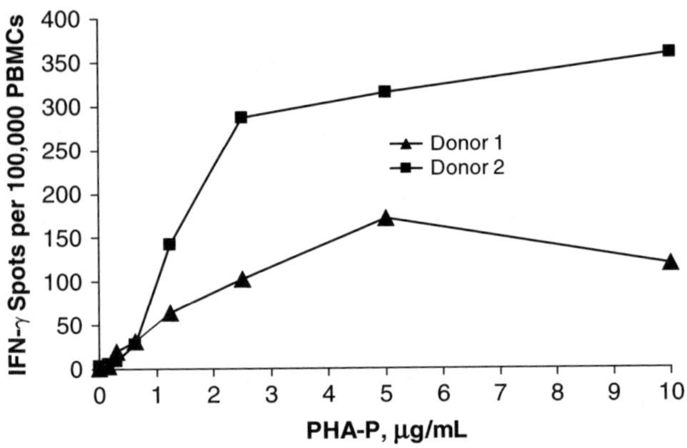

Figure 10.2 PHA-P (5 μg/mL) induced IFN-γ secretion by fresh and cryopreserved PBMCs. Data are mean ± SD of the results for fresh and frozen PBMCs

Optimal PHA-P concentration for measuring IFN-γ response

Freshly isolated PBMCs from two normal donor blood samples on the same day of blood draw were treated with varying concentrations of PHA-P ranging from 0.16 to 10.0 μg/mL. The cells were plated at a density of 100 000 cells/well. As shown in Figure 10.2, 5 μg/mL PHA-P was the optimal antigen concentration for *ex vivo* stimulation of IFN-γ responses from PBMCs.

Part 2: Studies with fresh frozen cells (refer to PBMCs isolated and cryopreserved on the same day of blood draw)

Cryopreserved PBMCs were studied to determine whether the findings for fresh PBMCs applied to fresh frozen cells. The results of these studies showed that cryopreserved and fresh PBMCs had similar IFN-γ responses.

Determination of optimal cell density for cryopreserved PBMCs

To determine the optimum number of cryopreserved PBMCs, fresh PBMCs, or PBMCs cryopreserved on the same day of blood collection from the same normal donors, were plated at cell densities ranging from 1000 to 100 000 cells and stimulated with PHA-P. The IFN-γ response in both fresh and cryopreserved PBMCs was comparable (Table 10.2). The cell density of 100 000 PBMCs produced an optimum response.

To determine whether cell densities >100 000 cells produce a better IFN-γ response, PBMCs isolated from on site collected blood from two normal donors and cryopreserved were plated at cell densities ranging from 5000 to 300 000 cells per well and stimulated with PHA-P. An optimal IFN-γ response of about 300 spots for both donors was observed at 100 000 cells

Table 10.2 Comparision of PHA-P induced IFN-γ secretion by PBMCs isolated (Fresh) and cryopreserved (Frozen) on the same day of blood draw[a]

Cell number	Fresh	Frozen	Mean	SD	%CV
1000	3	0	1.50	1.65	
2500	1	3	2.00	1.41	
5000	6	2	4.00	2.36	
10 000	12	11	11.67	0.47	4.04
25 000	48	48	48.17	0.24	0.49
50 000	166	108	137.33	41.01	29.86
100 000	316	309	312.33	5.19	1.66

[a] Data are expressed as number of IFN-γ spots per the indicated cell number.

Table 10.3 IFN-γ secretion by cryopreserved PBMCs[a]

Cell number	Donor 1	Donor 2
0	2	0
5000	0	0
10 000	0	0
25 000	3	27
50 000	67	91
75 000	162	229
100 000	256	332
300 000	515	508

[a] Data are expressed as number of IFN-γ spots at the indicated cell number

per well (Table 10.3). The spot number produced by a density of 100 000 normal cells per well is optimal because IFN-γ responses from patient samples that might be lower or higher than normal IFN-γ responses could be measured in the IFN-γ ELISPOT assay. Approximately 500 ELISPOTs were observed when the cell density was 300 000 cells/well. This response is not ideal because the IFN-γ response from patient PBMCs that produce higher amount of IFN-γ compared with normal PBMCs cannot be accurately determined. See the section “Limit of Quantification” later in this chapter for a discussion of the upper limit of quantification (ULOQ) for the IFN-γ ELISPOT assay.

PBMCs from several normal donors were examined before and after cryopreservation to determine the range of responses at the 100 000 cell density. A concentration of 5 μg/mL PHA-P was used to stimulate the cells. The IFN-γ response in normal donor PBMCs from six donors before and after cryopreservation is comparable with pooled%CV of 17.60 (Table 10.4). The number of IFN-γ spots ranged from 171 to 420 in fresh cells and 190 to 458 in cryopreserved cells.

Table 10.4 Comparision of IFN-γ secretion by fresh and cryopreserved PBMCs (isolated from same day of blood draw)[a]

Donor	0 hour fresh	0 hour cryo	Mean	SD	%CV
Donor 1	397	218	308	126.57	41.16
Donor 2	420	458	439	27.11	6.17
Donor 3	260	190	225	49.73	22.12
Donor 4	300	318	309	12.73	4.11
Donor 5	171	256	213	60.10	28.20
Donor 6	315	332	324	12.49	3.86
				pooled %CV	17.60

[a] Data are expressed as number of IFN-γ spots per 100 000 PBMCs.

Part 3: Validation studies with PBMCs isolated from shipped or laboratory stored blood samples

IFN-γ response from PBMCs isolated from laboratory stored peripheral blood samples

The IFN-γ response from PBMCs from a normal donor sample stored at room temperature for 24 and 48 hours stimulated with PHA-P was compared with the response of fresh PBMCs. The ability of PBMCs to secrete IFN-γ in response to PHA-P was substantially decreased by the duration of storage at room temperature. These results confirm an earlier observation that peptide mixes were more effective than whole protein antigens in stimulating responses from stored or shipped samples (Maecker et al, 2001). To determine if increasing the concentration of PHA-P enhances IFN-γ secretion by PBMCs from stored samples PBMCs stored for 24 hours at room temperature were stimulated with 5, 10, and 20 μg/mL of PHA-P. Higher concentrations of PHA-P did not increase the IFN-γ response from samples stored at room temperature (Figure 10.3).

IFN-γ response from overnight shipped blood samples

To simulate potential shipping conditions for clinical study samples, whole blood samples from four normal donors were shipped overnight at ambient temperature. PBMCs were isolated from an aliquot of these samples before shipment and tested for IFN-γ secretion before and after cryopreservation. PBMCs isolated from the overnight shipped samples immediately after receipt were also examined for IFN-γ secretion before cryopreservation and after cryopreservation and storage for 1 week. The IFN-γ response by PBMCs from overnight shipped samples that had not been cryopreserved was reduced compared with the response of fresh PBMCs in all

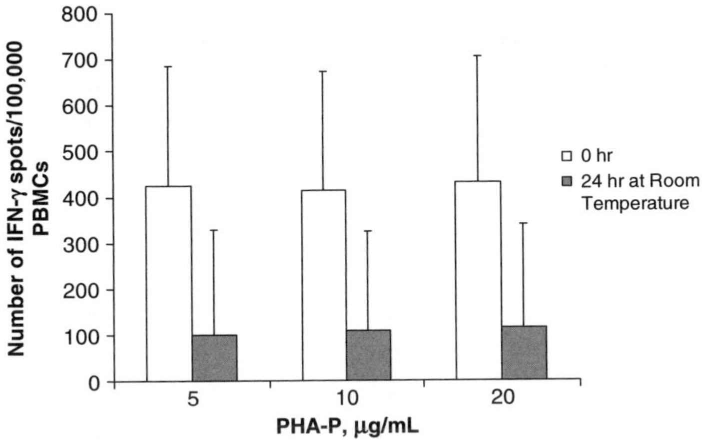

Figure 10.3 Effect of PHA-P concentration on IFN secretion by cryopreserved PBMCs prepared from samples stored at room temperature

Table 10.5 Mean IFNγ secretion by PBMCs stimulated by PHA-P before and after overnight shipment prior to cryopreservation[a,b]

Donor	Fresh 0 hour	Fresh 24 hours
1	397	98
2	420	77
3	260	94
4	300	58

[a] Values are the number of IFN-γ spots per 100 000 PBMCs.
[b] PBMC indicates peripheral blood mononuclear cell; PHA-P, phytohemagglutinin-P; fresh, not cryopreserved.

four donors (Table 10.5). Similarly, cryopreserved PBMCs from overnight shipped samples had a reduced IFN-γ response compared with cryopreserved cells from fresh PBMCs (Table 10.6). The reduced IFN-γ response appeared to be due to overnight shipment and not cryopreservation because freshly isolated PBMCs and cryopreserved PBMCs from overnight shipped samples collected from the same four normal donors had similar IFN-γ responses to PHA-P (Table 10.7). Therefore, all clinical samples will be exposed to the same shipping and cryopreservation conditions to ensure consistency of the data obtained.

To determine if samples received after 2 days of shipment respond to stimulation with PHA-P, the IFN-γ response of fresh PBMCs and those cryopreserved before and after 1 or 2 days of shipment was assessed. Cryopreserved PBMCs received 1 day after blood collection showed a reduced response compared to PBMCs isolated from samples before shipment (Table 10.8). The IFN-γ response was further reduced in PBMCs

Table 10.6 Mean IFNγ secretion by cryopreserved PBMCs stimulated by PHA-P from overnight shipped samples[a,b]

Donor	Cryopreserved 0 hour	Cryopreserved 24 hours
1	218	154
2	458	65
3	190	105
4	318	48
5	90	32
6	19	10
7	101	19
8	437	184
9	393	151

[a] Values are the number of IFN-γ spots per 100 000 PBMCs.
[b] PBMC indicates peripheral blood mononuclear cell; PHA-P, phytohemagglutinin-P.

Table 10.7 Mean IFNγ secretion by cryopreserved PBMCs stimulated by PHA-P from overnight shipped samples before and after cryopreservation[a,b]

Donor	Fresh 24 hours	Cryopreserved 24 hours
1	98	154
2	77	65
3	94	105
4	58	48
5	171	256
6	315	332

[a] Values are the number of IFN-γ spots per 100 000 PBMCs.
[b] PBMC indicates peripheral blood mononuclear cell; PHA-P, phytohemagglutinin-P.

isolated from samples received on the second day after blood collection (Table 10.8). The reduced response of shipped samples was not attributed to reduced viability at 24 or 48 hours after sample collection compared with fresh PBMCs. The mean viability of freshly isolated PBMCs from fresh blood samples, overnight shipped samples, and 2-day shipped samples was 98.5%, 92.2% and 92.7%, respectively. Similar mean viability results were obtained for PBMCs isolated and cryopreserved from fresh blood samples (97.9% viable), overnight shipped samples (96.5% viable). Cell viability was reduced to a mean of 86.3% in PBMCs cryopreserved from samples in shipment for 2 days. Because only viable cells are used in the assay, the reduced functionality of the PBMCs reflected in the weaker IFN-γ response can be attributed to a lower number and functionality of antigen-presenting cells preserved after shipment and freezing (Kumar and Satchidanandam, 2000).

Table 10.8 PHA-P induced IFN-γ secretion by cryopreserved PBMCs from shipped samples[a,b]

Donor	Before shipment	After sample receipt	
		Day 1	Day 2
Mean Donor 1	450	58	28
%Change (before and after)		87.11	93.78
Mean Donor 2	633	167	112
%Change (before and after)		73.62	82.31
Mean Donor 3	377	210	55
%Change (before and after)		44.30	85.41
Mean Donor 4	169	81	11
%Change (before and after)		52.07	93.49
Mean Donor 5	184	30	1
%Change (before and after)		83.65	99.27
Mean Donor 6	16	24	0
%Change (before and after)		−53.19	100.00
Mean Donor 7	192	58	21
%Change (before and after)		69.62	89.06
Mean Donor 8	81	67	8
%Change (before and after)		17.70	90.12
Mean Donor 9	69	5	0
%Change (before and after)		92.79	100.00
Mean Donor 10	471	63	5
%Change (before and after)		86.63	98.87
Mean Donor 11	352	60	5
%Change (before and after)		83.05	98.67
Mean Donor 12	237	133	13
%Change (before and after)		43.80	94.51
Mean Donor 13	653	315	26
%Change (before and after)		51.81	96.07
Mean Donor 14	503	211	63
%Change (before and after)		58.18	87.58
Mean Donor 15	574	392	17
%Change (before and after)		31.61	97.04
Mean Donor 16	406	131	38
%Change (before and after)		67.73	90.64
Mean Donor 17	570	278	55
%Change (before and after)		51.20	90.35
Mean Donor 18	312	148	6
%Change (before and after)		52.46	97.97

Table 10.8 (Continued)

Donor	Before shipment	After sample receipt	
		Day 1	Day 2
Mean Donor 19	450	71	17
%Change (before and after)		84.14	96.22
Mean Donor 20	356	32	2
%Change (before and after)		90.91	99.58
Mean Donor 21	236	67	7
%Change (before and after)		71.75	97.25
Mean Donor 22	552	195	27
%Change (before and after)		64.63	95.17
Mean Donor 23	446	64	13
%Change (before and after)		85.74	97.20

[a] Values are the number of IFN-γ spots per 100 000 PBMCs.
[b] PHA-P indicates phytohemagglutinin-P; PBMC indicates peripheral blood mononuclear cell.

The IFN-γ response of cryopreserved PBMCs from overnight shipped samples can be enhanced by increasing the cell density from 100 000 cells per well used in the experiments in the preceding two paragraphs to 200 000 or 300 000 cells per well (Table 10.9 and Figure 10.4). Dilution of cell numbers in the assay well reduces both responding cells and antigen-presenting cells, which can affect cell contact between the two populations needed for cytokine responses (Smith et al, 2001). Low to high responding cryopreserved PBMCs from overnight shipped samples can be tested at a density of 200 000 cells per well, whereas a density of 300 000 cells per well can be used to detect very low frequency responses, but not high responders. If high responders are tested at a density of 300 000 cells per well, it is probable that spot number could reach the HLOQ of a mean of 700 spots (See the "Limit of quantification" section later in this chapter).

Table 10.9 Effect of cell density on IFN-γ secretion by cryopreserved PBMCs from overnight shipped samples[a,b]

Cell number	Donor 1	Donor 2	Donor 3	Donor 5	Donor 6	Donor 7	Donor 9	Donor 10	Donor 11	Donor 12	Donor 13
100 000	49	19	16	24	16	16	24	16	11	10	26
200 000	107	216	58	67	43	58	67	43	63	60	133
300 000	195	401	107	140	67	107	140	67	146	202	254

[a] Values are the number of IFN-γ spots per 100 000 PBMCs.
[b] PBMC indicates peripheral blood mononuclear cell.

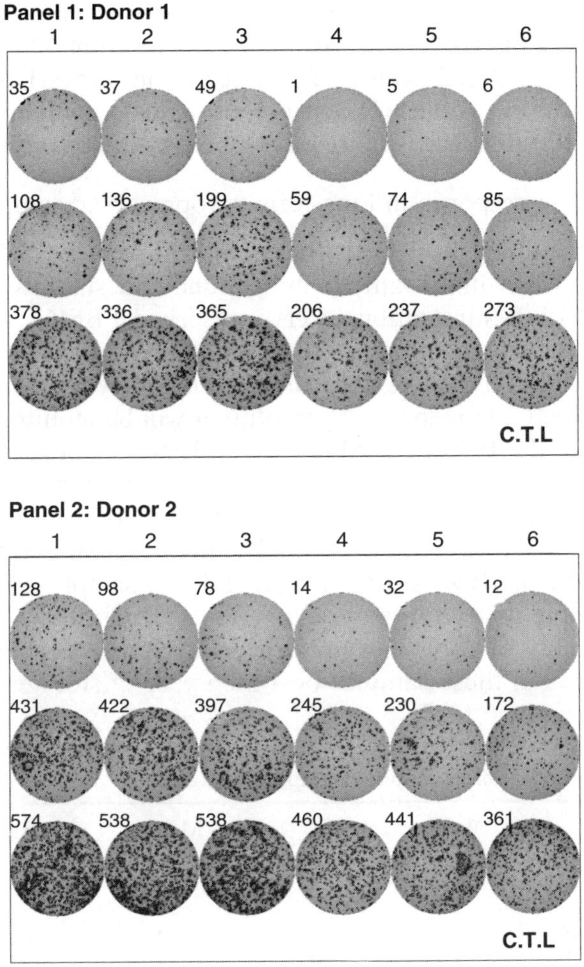

Figure 10.4 Representative responses for PBMCs from two normal donors plated at cell densities of 100 000, 200 000, and 300 000 cells per well (Rows 1, 2, and 3, respectively, in Panels 1 and 2). Columns 1 to 3 in each panel are IFN-γ responses by PBMCs cryopreserved on the day that the whole blood sample was collected. Columns 4 to 6 in each panel are IFN-γ responses by PBMCs cryopreserved the day after collection of the whole blood samples

Intersubject variability in fresh and cryopreserved PBMCs

Fresh PBMCs and PBMCs cryopreserved on the day of sample collection or from overnight shipped samples were stimulated with PHA-P. All PBMCs were from normal donors. A wide range of IFN-γ responses to stimulation with PHA-P were observed in fresh PBMCs and PBMCs cryopreserved on the day of sample collection or from overnight shipped samples. The %CVs ranged from 64.39 to 76.40 with mean spots of 337 ± 64.39. This mean spot value is similar to the previously reported mean spot value of 328 ± 66 for IFN-γ secretion in response to PHA-P stimulation in PBMCs from 52 normal subjects (Hagiwara et al, 1995). Because IFN-γ secretion varies between donors, the IFN-γ secretion profile of each patient will be compared to baseline values in clinical studies (pre-treatment vs. post-treatment comparison).

Stability of IFN-γ response in long-term cryopreserved samples

Because pre-treatment and post-treatment comparisons will be made in clinical studies, it is important to understand the stability of the IFN-γ response to PHA-P stimulation in cryopreserved PBMCs with long-term storage. The duration of the long-term stability studies should match the anticipated duration of the planned clinical study in which changes in IFN-γ secretion by PBMCs in response to treatment will be monitored. As part of the validation of the IFN-γ ELISPOT assay, IFN-γ responses from PBMCs isolated and cryopreserved from overnight shipped samples and stimulated with PHA-P were monitored over 9 weeks. The IFN-γ response in long term stored PBMCs was well preserved with %CV ranging from 13.46 to 28.97 (Table 10.10). PBMCs from a normal donor sample cryopreserved on the day of sample collection were examined periodically over a 16-month period for stability of IFN-γ response to PHA-P. It was found that IFN-γ could be induced in these samples beyond a year of storage (Table 10.11).

Table 10.10 Effect of long-term cryopreservation on the IFN-γ response of PBMCs cryopreserved from overnight shipped samples[a,b]

Donor	Weeks			Mean	SD	%CV
	0	3	9			
1	258	169	218	194	34.65	17.91
2	366	450	297	374	108.19	28.97
2	ND	633	523	578	77.78	13.46

[a] Values are the number of IFN-γ spots per 200 000 PBMCs.
[b] PBMC indicates peripheral blood mononuclear cell.

Table 10.11 Effect of long-term cryopreservation on the IFN-γ response by PBMCs cryopreserved on the day of sample collection[a,b]

Donor	Months							Mean	SD	%CV
	0	7	9	12	13	15	16			
1	332			143	93	126	275	194	103.72	53.52
2	659	387	556					534	137.48	25.74

[a] Values are the number of IFN-γ spots per 100 000 PBMCs.
[b] PBMC indicates peripheral blood mononuclear cell.

Thus PBMCs from one normal donor stimulated with a polyclonal activator can be used as a positive control in all ELISPOT assays conducted for a particular trial. Though this use of a normal control confirms that the assay was performed accurately, it does not provide any information on the integrity of cryopreserved clinical samples. The integrity of the clinical samples is provided to some extent by cell viability assessments.

Part 4: Assay precision and other validation parameters using PBMCs isolated and cryopreserved from overnight shipped blood samples

Assay precision in the IFN-γ ELISPOT assay with PHA-P stimulation

Intra-assay variability

Cryopreserved PBMCs from overnight shipped four normal human whole blood samples were used to examine intra-assay variability of the IFN-γ ELISPOT assay. The samples were assayed in triplicate wells in two different assay plates to evaluate intra-assay variability of the assay. The determination of IFN-γ ELISPOTs in triplicate wells was highly precise (Tables 12a and 12b). A pooled SD of 28.11 and %CV of 11.93 in the first plate (Table 10.12) and a pooled SD of 26.29 and %CV of 10.53 in the second plate (Table 10.13) were obtained for 14 donors. These results indicate that intra-assay variability is acceptably low and justifies the analysis of clinical samples in duplicate when samples are limited because of low cell count or poor recovery of the cells after thawing.

Table 10.12 Intra-assay variability in the IFN-γ ELISPOT assay (plate A)[a,b]

Donor	Replicate 1	Replicate 2	Replicate 3	Mean	SD	%CV
1	525	505	ND	515	14.14	2.75
2	254	272	248	258	12.49	4.84
3	94	118	109	107	12.12	11.33
4	447	357	295	366	76.43	20.86
5	514	549	543	535	18.72	3.50
6	164	190	181	178	13.20	7.40
7	50	72	58	60	11.14	18.56
8	402	345	301	349	50.64	14.50
9	420	395	395	403	14.43	3.58
10	210	ND	317	264	75.66	28.71
12	443	469	434	449	18.18	4.05
13	107	173	204	161	49.54	30.71
14	316	311	293	307	12.10	3.94
Pooled					28.11	11.93

[a] Values are the number of IFN-γ spots.
[b] PBMC indicates peripheral blood mononuclear cell, ND, not determined.

Table 10.13 Intra-assay variability in the IFN-γ ELISPOT assay (plate B)[a,b]

Donor	Replicate 1	Replicate 2	Replicate 3	Mean	SD	%CV
1	451	539	516	502	45.64	9.09
2	242	257	263	254	10.82	4.26
3	135	144	114	131	15.39	11.75
4	405	321	342	356	43.71	12.28
5	442	443	425	437	10.12	2.32
6	162	122	178	154	28.84	18.73
7	52	68	60	60	8.00	13.33
8	295	283	226	268	36.86	13.76
9	279	350	ND	315	50.20	15.96
10	243	178	ND	211	45.96	21.83
12	352	387	438	392	43.25	11.02
13	131	131	ND	131	0.00	0.00
14	293	263	ND	278	21.21	7.63
Pooled					26.29	10.53

[a] Values are the number of IFN-γ spots.
[b] PBMC indicates peripheral blood mononuclear cell, ND, not determined.

Inter-assay precision

To evaluate inter-assay precision, cryopreserved PBMCs prepared from samples shipped overnight to mimic clinical sample delivery were plated into two different plates on the same day. The same sample of PBMCs was used in both plates. Inter-assay variability of the two assays performed on the same day with cryopreserved PBMCs from 20 normal donors isolated from overnight shipped samples showed low interplate variability (Table 10.14). The pooled SD and %CV were 59.44 and 18.87, respectively.

Inter-assay variability was also assessed in assays performed on three consecutive days using PBMCs from five donors that had been cryopreserved from overnight shipped samples and cryopreserved PBMCs from two donors that had been stored for 17 months to evaluate the potential usefulness of the cells as positive controls during the analysis of clinical samples. The within-day variability from this experiment can be viewed as inter-assay results because different cryovials PBMC were used in the ELISPOT assay on each day. The cryovials were from the same lot of frozen cells. The results for the within-day variability experiments suggested that interday variability of the IFN-γ ELISPOT assay was higher than intra-assay or interplate variability with pooled %CV of 25.10 for PBMCs cryopreserved from overnight shipped samples (Table 10.15) and 37.45 for cells stored for 17 months. Part of this inter-day variability is likely due to the use of a different PBMC cryovial on each day. Even though the cryovials were from the same lot, sample-to-sample variability in cell viability and functionality of the cells could explain the findings. Therefore, all samples from a clinical

Table 10.14 Inter-assay variability of the IFN-γ ELISPOT assay in PBMCs cryopreserved from overnight shipped samples[a,b]

Donor	Plate A	Plate B	Plates A and B		
	Mean		Mean	SD	%CV
1	515	502	509	9.19	1.81
2	258	254	256	2.83	1.10
3	107	131	119	16.97	14.26
4	366	356	361	7.07	1.96
5	566	558	562	5.66	1.01
6	502	272	387	162.63	42.02
7	356	147	252	147.79	58.76
8	337	359	348	15.32	4.40
9	320	405	363	59.63	16.45
10	216	318	267	72.48	27.17
11	272	437	355	116.67	32.91
12	147	268	208	85.56	41.23
13	359	343	351	10.84	3.09
14	405	416	410	7.78	1.90
15	318	471	394	107.95	27.38
16	502	437	470	45.96	9.79
17	356	268	312	62.23	19.94
18	337	343	340	4.48	1.32
19	320	416	368	67.41	18.32
20	216	471	343	180.43	52.59
Pooled				59.44	18.87

[a] Values are the number of IFN-γ spots per 200 000 PBMCs.
[b] PBMC indicates peripheral blood mononuclear cell.

Table 10.15 Inter-assay (inter-day) variability of the IFN-γ ELISPOT assay in PBMCs cryopreserved from overnight shipped samples[a,b]

Donor	Day 1	Day 2	Day 3	Mean	SD	%CV
1	502	272	437	404	118.57	29.37
2	356	147	268	257	104.93	40.83
3	337	359	343	346	11.14	3.22
4	320	405	416	380	52.16	13.72
5	216	318	471	335	128.40	38.36
Pooled					83.04	25.10

[a] Values are the number of IFN-γ spots per 200 000 PBMCs.
[b] PBMC indicates peripheral blood mononuclear cell.

Table 10.16 Inter-operator variability of the IFN-γ ELISPOT assay[a,b]

Donor	Operator 1	Operator 2	Mean	SD	%CV
	Mean				
1	535	437	486	69.30	14.26
2	178	154	166	16.97	10.22
3	60	60	60	0.00	0.00
4	349	268	309	57.28	18.57
5	403	315	359	62.81	17.50
6	264	211	237	37.48	15.81
7	449	392	421	39.83	9.47
8	161	131	146	21.45	14.67
9	307	278	292	20.27	6.93
10	120	148	134	20.03	14.93

[a] Values are the number of IFN-γ spots per 200 000 PBMCs.
[b] PBMC indicates peripheral blood mononuclear cell.

study subject should be tested on the same day. The results from long-term cryopreserved samples from two donors suggest that these samples can be used as assay controls when clinical samples are assayed on different days.

Inter-operator precision

Inter-operator precision was evaluated when two trained operators tested cryopreserved PBMCs isolated from overnight shipped samples from ten normal donors in the IFN-γ ELISPOT assay on the same day (Table 10.16). PHA-P was used as the activating agent. Inter-operator variability in the IFN-γ ELISPOT assay was low with %CV range of 0.00 to 18.57 for the two operators for results when cells were stimulated with PHA-P. The results for the two operators for the negative control or unstimulated cells were also similar. The intra-assay variability of the assays performed by the two operators had an SD of 27.83 and 25.25 with %CV of 12.72 and 11.00 for Operator 1 and 2, respectively.

Effect of different lots of reagents and assay components on PHA-P induced IFN-γ responses

The following experiments were conducted to ascertain how to maintain a validated state as described in the preceding chapter in the event of changes in lots and sources of reagents and assay plates.

Effect of different lots of PHA-P

Two different lots of PHA-P were used to examine the effect or variability in different lots of PHA-P. Cryopreserved PBMCS from four overnight shipped samples were plated in triplicate wells and stimulated with 5 μg/mL PHA-P. The PHA-P was from two different lots and both lots were tested on the same plate. The IFN-γ response induced by both lots of PHA-P was comparable in all four donors with a %CV range of 8.97 to 25.19 between the two lots (data not shown). Thus, new lots of PHA-P can be used during a clinical study without re-titrating the concentration of the mitogen or cell density. Nevertheless, all samples from one subject collected at different time points should be tested using the same lot of mitogen.

Effect of different lots of assay plates

Cryopreserved PHA-P stimulated PBMCs from four overnight shipped samples were plated in triplicate wells in two different lots of assay plates (MultiScreenTM–IP with PVDF membrane) to examine the effect the potential effect of different lots of assay plates on the IFN-γ ELISPOT assay. Comparable results were obtained in both ELISPOT assay plates with all four donors with a %CV range of 0.95 to 13.09. The negligible background response in wells incubated without stimulation was also similar in both assay plates (data not shown). Therefore, new lots of assay plates can be used during a clinical study without re-titrating the concentration of mitogen or cell density. However, all samples from one subject collected at different time points during a clinical study should be tested using the same lot of assay plates to minimize variability in the assay results.

Effect of assay plates from different sources

Assay components, such as IFN-γ antibody pairs (capture and detection antibodies) and streptavidin horseradish peroxidase, are routinely purchased from BD Biosciences-Pharmingen (San Diego, CA) and assay plates from Millipore Corporation (Billerica, MA) as individual items. When assay plates or antibody pairs are not available from these vendors, it is necessary to use the human IFN-γ ELISPOT set from BD Biosciences-Pharmingen (San Diego, CA). The IFN-γ ELISPOT set includes ten plates from Millipore and the same clones of capture and detection antibodies as the antibody pairs as well as streptavidin horseradish peroxidase. All reagents are packaged as a set to perform ten-plate assays. To determine whether the routinely used assay plates and antibody pairs or the IFN-γ ELISPOT assay set produce different results, cryopreserved PBMCs from an overnight shipped normal donor sample were plated in triplicate wells into assay plates from two different sources, a MultiScreenTM–IP and a BD ELISPOT set plate, and the PHA-P induced IFN-γ response was measured using individual components or reagents from the ELISPOT set. Comparable results were obtained in both ELISPOT assay plates with %CV range of 21.02 between the plates (data not shown). Therefore, the IFN-γ ELISPOT assay can be performed using reagents purchased as individual components or as a set without re-titrating the mitogen concentration or cell density. It is more cost effective to use individual components and their use is recommended for testing clinical samples. All samples from one subject that have been collected at different time points should be tested using same reagents to minimize possible variability that may contribute to the variability of the results.

Stability of capture antibody-coated plates

To determine the reproducibility of IFN-γ response in plates that have been coated with anti-cytokine antibody and stored at 4 °C for different lengths of time, PBMCs from normal donors cryopreserved the same day (100 000 cells per well) or after overnight shipment (200 000 or 300 000 cells per well) were added to assay plates that had been coated with anticytokine antibody one to three days prior to the assay. The SD and %CV of the PHA-P induced IFN-γ response ranged from 0.24 to 11.79 and 3.75 to 6.32, respectively, in wells coated 1 or 2 days before the assay for all cell densities (data shown). Comparable IFN- γspots were observed in the wells. There was higher variability in the IFN-γ response in wells coated three days before the assay the assay with an SD of 4.24 to 35.12 and %CV of 20.35 to 27.92 for all cell densities. Only plates coated no more than 2 days before the assay date can be used for the IFN-γ ELISPOT assay.

10.4 Limit of quantification

The lower limit of quantification (LLOQ) summarized in Table 10.17 for PHA-P, PMA + ionomycin, and SEB has been determined to be a mean spot number of 10, based on the IFN-γ responses in different assays. High %CV values tend to be obtained when spot numbers in triplicate wells are low. To capture the low frequency responses mainly to a specific antigen, the LLOQ has been set at a mean spot number of 10 even though less variability is observed when spot numbers are >20. If the %CV is >30% at a LLOQ of 10, samples should be re-assayed at using a higher cell density.

Based on the number and morphology of the spots, The ULOQ for the IFN-γ ELISPOT assay is 700 spots per well with assay precision of ≤30% CV between triplicate wells based on the number and the morphology of the spots (Table 10.17). Responses resulting in spot numbers >600 are usually an estimate based on mean spot sizes and therefore not accurate.

Table 10.17 Lower limit and upper limit of quantification for the IFN-γ ELISPOT assay

	Stimulation	Cells per well	SFC per well[a]	SFC/PMC
LLOQ	Unstimulated	100 000	<10	<100
	PHA-P	100 000	<10	<100
	PHA-P	200 000	<10	<50
	PMA + ionomycin	10 000	<10	<1 000
	SEB	100 000	<10	<100
	SEB	200 000	<10	<50
ULOQ	Unstimulated	100 000	>600	>6 000
	PHA-P	100 000	>600	>6 000
	PHA-P	200 000	>600	>3 000
	PMA + ionomycin	10 000	>600	>60 000
	SEB	100 000	>600	>6 000
	SEB	200 000	>600	>3 000

[a] SFC indicates spot forming cells; PMC, per million cells; LLOQ, lower limit of quantification; ULOQ, upper limit of quantification; PHA-P, phytohemagglutinin-P; PMA, phorbol-12-myristate 13-acetate; SEB, staphylococcal enterotoxin B.

10.5 Assay conditions for measuring antigen-induced IFN-γ responses

The following IFN-γ ELISPOT assay conditions are recommended for clinical sample based on the results of the validations studies using the stimulating agents, PHA-P, PMA + ionomycin, and SEB.

- Each sample should be tested in triplicate.
- PBMCs can be analyzed in duplicate if limited sample is available.
- ELISPOT assay plates should be coated with IFN-γ capture antibodies not more than one day prior to the assay.
- All samples collected at different time points from a clinical study subject should be tested on the same day and in a single plate to avoid interday variability.
- Samples collected at all time points from a clinical study subject should be plated on the same day by the same operator to minimize variability.
- All samples from a clinical study subject collected at different time points should be analyzed using the same lot and source of plates, as well as other assay reagents to avoid interplate variability because of different lots of plates.
- As an assay and reagent control for each assay plate, a pre-tested normal donor sample (cryopreserved as a large batch of aliquots) should be included in each plate.

- Each plate should include a positive control sample plated in triplicate with or without stimulation with PHA-P, PMA + ionomycin, and SEB. The cell density of the positive control sample is based on pre-tested conditions for each stimulating agent.
- Each clinical sample should be tested with and without the stimulating agents in the same plate, PHA, PMA + ionomycin, and/or SEB, to detect any spontaneous responses.

10.6 Clinical sample acceptance criteria

The following acceptance criteria for clinical samples for the IFN-γ ELISPOT assay are based on the results of the validations studies using the stimulating agents, PHA-P, PMA + ionomycin, and SEB.

- Blood samples received the same day or the day after collection can be accepted for the isolation and cryopreservation of PBMCs.
- Blood samples received 2 days after collection are excluded from ELISPOT analysis.
- Cryopreserved and thawed PBMCs with percent viability of <70 ascertained by Guava PCA system should not be included in ELISPOT testing.

10.7 Summary

An accurate and precise ELISPOT assay for the quantitative determination of IFN-γ secretion by cryopreserved and stimulated human PBMCs isolated from peripheral whole blood samples shipped the same day or the next day has been successfully developed and validated. The IFN-γ ELISPOT assay conditions were optimized for cell number, concentration of stimuli, and the cell density of cryopreserved PBMCs to produce an optimal and reproducible IFN-γ response using positive controls. The number of spots from PBMCs isolated the same day or the day after sample collection was not affected by cryopreservation and long-term storage for more than 1 year. Cryopreserved PBMCs from overnight shipped samples were induced to secrete IFN-γ with high reproducibility in triplicate wells with low intra-assay variability. Even though interday (inter-assay) variability was higher than intra-assay variability, the accuracy of the ELISPOT assay results should not be affected if samples from all time points from a clinical study subject are analyzed on the same day. Multiple plates can be used to assay samples from a single subject, if needed, because interplate variability was minimal. Inter-operator variability was also low confirming the precision of the assay procedure. The final results from patient samples are expressed as mean SFC or NCS per million PBMCs and as percent change from baseline or pre-treatment samples.

References

Bennouna J, Hildesheim A, Chikamatsu K, Gooding W, Storkus WJ and Whiteside TL (2002). Application of IL-5 ELISPOT assays to quantification of antigen-specific T helper responses. *J Immunol Methods*, **261,** 145–156.

Boehm U, Klamp T, Groot M and Howard JC (1997). Cellular responses to interferon-gamma. *Annu Rev Immunol*, **15,** 749–795.

Clay TM, Hobeika AC, Mosca PJ, Lyerly HK and Morse MA (2001). Assays for monitoring cellular immune responses to active immunotherapy of cancer. *Clin Cancer Res*, **7,** 1127–1135.

Currier JR, Kuta EG, Turk E, Earhart LB, Loomis-Price L, Janetzki S, Ferrari G, Birx DL and Cox JH (2002). A panel of MHC class I restricted viral peptides for use as a quality control for vaccine trial ELISPOT assays. *J Immunol Methods*, **260**, 157–172.

Farina C, Then Bergh F, Albrecht H, Meinl E, Yassouridis A, Neuhaus O and Hohlfeld R (2001). Treatment of multiple sclerosis with Copaxone (COP): Elispot assay detects COP-induced interleukin-4 and interferon-gamma response in blood cells. *Brain*, **124**, 705–719.

Hagiwara E, Abbasi F, Mor G, Ishigatsubo Y and Klinman DM (1995). Phenotype and frequency of cells secreting IL-2, IL-4, IL-6, IL-10, IFN and TNF-alpha in human peripheral blood. *Cytokine*, 7, 815–822.

Hedman M, Ludvigsson J and Faresjo MK (2006). Nicotinamide reduces high secretion of IFN-gamma in high-risk relatives even though it does not prevent type 1 diabetes. *J Interferon Cytokine Res*, **26**, 207–213.

Helms T, Boehm BO, Asaad RJ, Trezza RP, Lehmann PV and Tary-Lehmann M (2000). Direct visualization of cytokine-producing recall antigen-specific CD4 memory T cells in healthy individuals and HIV patients. *J Immunol*, **164**, 3723–3732.

Jansson A, Ernerudh J, Kvarnstrom M, Ekerfelt C and Vrethem M (2003). Elispot assay detection of cytokine secretion in multiple sclerosis patients treated with interferon-beta1a or glatiramer acetate compared with untreated patients. *Mult Scler*, **9**, 440–445.

Johnson HM, Russell JK and Pontzer CH (1991). Staphylococcal enterotoxin microbial superantigens. *Faseb J*, 5, 2706–2712.

Kotzin BL, Leung DY, Kappler J and Marrack P (1993). Superantigens and their potential role in human disease. *Adv Immunol*, 54, 99–166.

Kreher CR, Dittrich MT, Guerkov R, Boehm BO and Tary-Lehmann M (2003). CD4+ and CD8+ cells in cryopreserved human PBMC maintain full functionality in cytokine ELISPOT assays. *J Immunol Methods*, 278, 79–93.

Kumar P and Satchidanandam V (2000). Ethyleneglycol-bis-(beta-aminoethylether) tetraacetate as a blood anticoagulant: preservation of antigen-presenting cell function and antigen-specific proliferative response of peripheral blood mononuclear cells from stored blood. *Clin Diagn Lab Immunol*, 7, 578–583.

Letsch A and Scheibenbogen C (2003). Quantification and characterization of specific T-cells by antigen-specific cytokine production using ELISPOT assay or intracellular cytokine staining. *Methods*, **31**, 143–149.

Maecker HT, Dunn HS, Suni MA, Khatamzas E, Pitcher CJ, Bunde T, Persaud N, Trigona W, Fu TM, Sinclair E, Bredt BM, Mccune JM, Maino VC, Kern F and Picker LJ (2001). Use of overlapping peptide mixtures as antigens for cytokine flow cytometry. *J Immunol Methods*, 255, 27–40.

Maino VC (1998). Rapid assessment of antigen induced cytokine expression in memory T cells by flow cytometry. *Vet Immunol Immunopathol*, 63, 199–207.

Marrack P and Kappler J (1990). The staphylococcal enterotoxins and their relatives. *Science*, **248**, 705–711.

Mehlis S and Gordon KB (2004). From laboratory to clinic: rationale for biologic therapy. *Dermatol Clin*, 22, 371–377, vii–viii.

Navikas V and Link H (1996). Review: cytokines and the pathogenesis of multiple sclerosis. *J Neurosci Res*, **45**, 322–333.

Okamoto Y, Gotoh Y, Shiraishi H and Nishida M (2004). A human dual-color enzyme-linked immunospot assay for simultaneous detection of interleukin 2- and interleukin 4-secreting cells. *Int Immunopharmacol*, **4**, 149–156.

Pala P, Hussell T and Openshaw PJ (2000). Flow cytometric measurement of intracellular cytokines. *J Immunol Methods*, **243**, 107–124.

Pelfrey CM, Rudick RA, Cotleur AC, Lee JC, Tary-Lehmann M and Lehmann PV (2000). Quantification of self-recognition in multiple sclerosis by single-cell analysis of cytokine production. *J Immunol*, **165**, 1641–1651.

Ronnelid J, Berg L, Rogberg S, Nilsson A, Albertsson K and Klareskog L (1998). Production of T-cell cytokines at the single-cell level in patients with inflammatory arthritides: enhanced activity in synovial fluid compared to blood. *Br J Rheumatol*, **37**, 7–14.

Smith JG, Liu X, Kaufhold RM, Clair J and Caulfield MJ (2001). Development and validation of a gamma interferon ELISPOT assay for quantitation of cellular immune responses to varicella-zoster virus. *Clin Diagn Lab Immunol*, **8**, 871–879.

Tominaga K, Yoshimoto T, Torigoe K, Kurimoto M, Matsui K, Hada T, Okamura H and Nakanishi K (2000). IL-12 synergizes with IL-18 or IL-1beta for IFN-gamma production from human T cells. *Int Immunol*, **12**, 151–160.

Zollner TM, Wichelhaus TA, Hartung A, Von Mallinckrodt C, Wagner TO, Brade V and Kaufmann R (2000). Colonization with superantigen-producing Staphylococcus aureus is associated with increased severity of atopic dermatitis. *Clin Exp Allergy*, **30**, 994–1000.

11 IL-5 ELISPOT assay validation

Manjula Reddy[1], Jackson Wong, Hugh Davis, Charles Pendley and Uma Prabhakar

Centocor Research and Development, Inc., 145 King of Prussia Road, Radnor, PA 19087, USA

11.1 Introduction

The two subsets of helper T cells (T_H), T_H1 and T_H2, may be distinguished based on their cytokine secretion profile. T_H1 cells are reported to secrete interferon-gamma (IFN-γ), interleukin (IL)-12, and IL-2, whereas IL-4, IL-5, IL-6, and IL-10 and/or IL-13 are associated with T_H2 cells. T_H1 cells are primarily involved with helper activity for cell-mediated responses and T_H2 cells are involved with the process of B-cell differentiation and antibody production (Bennouna et al, 2002). It has become increasingly important to monitor patients with cancer or infectious diseases receiving therapeutic vaccines for assessment of changes in the frequency of circulating T_H1-type and T_H2-type cells to evaluate the magnitude of the T-cell response to vaccines and relate this to clinical outcome.

[1] To whom correspondence should be addressed.

Validation of Cell-Based Assays in the GLP Setting: A Practical Guide Editors, Uma Prabhakar and Marian Kelley

The enzyme-linked immunospot or ELISPOT assay is emerging as a tool to evaluate T cell cytokine responses in clinical studies (Jansson et al, 2003; Pelfrey et al, 2000; Ronnelid et al, 1998). This assay is the most sensitive and robust technique for direct *ex vivo* monitoring of cytokine secreting treated or untreated peripheral blood mononuclear cells (PBMCs) when computerized image analysis is used (Clay et al, 2001; Kreher et al, 2003). In the ELISPOT assay, secreted cytokines immediately adhere to the solid phase of the ELISPOT assay plate and are not consumed by other cells in the culture or affected by simultaneously produced antagonists. Thus, the ELISPOT assay measures the net production of cytokines more stringently than other methods, such as enzyme-linked immunosorbent assays. It also allows evaluation of cryopreserved human lymphocytes without loss of functional responses (Bennouna et al, 2002; Kreher et al, 2003).

IL-5, a T_H2-type cytokine, is known to be involved in the regulation of humoral responses. An IL-5 ELISPOT assay has been used to measure proteolipid protein-specific T cells in patients with multiple sclerosis (MS) (Correale and Fiol, 2004; Pelfrey et al, 2000). In an experimental model of MS, induction of T_H2 responses appeared to enhance autoimmune reactions and promote humoral autoimmunity (Genain et al, 1996). IL-4 secretion has also been monitored in patients with MS and rheumatoid arthritis using IL-4 ELISPOT assay (Kanik et al, 1998; Kvarnstrom et al, 2004; Okamoto et al, 2004; Ronnelid et al, 1998). IL-5 was selected for the evaluation of T_H2 immunologic responses in clinical trial samples rather than IL-4 or IL-10 because IL-5 production is confined to T cells, while IL-4 and IL-10 can be produced by antigen-presenting cells and cause high background and bystander activity (Pelfrey et al, 2000). IL-5 selection was also based on better responses (secretion and reproducibility) observed for IL-5 in normal volunteers compared with IL-4.

Validation of the IL-5 ELISPOT assay for any positive control antigens of interest should include the identification of the appropriate cell number for optimal induction of IL-5 by both fresh peripheral blood mononuclear cells (PBMCs) and after cryopreservation soon after blood collection, or after overnight shipment of blood samples along with evaluation of inter-subject variability and assay precision (intra-assay, inter-assay, and inter-operator). These concepts are described in more detail in Chapter 10. The validation of the IL-5 ELISPOT assay using the positive control antigen phytohemagglutinin-P (PHA-P) as an example is described in this chapter.

11.2 Methods

Peripheral blood mononuclear cells

Because functional preservation of PBMCs is vital for a successful cytokine response, proper cryopreservation and thawing methods are of importance, especially for clinical samples that will be tested at a future date. To preserve the function and cytokine responsiveness of PBMCs, cells isolated from whole blood are cryopreserved and stored frozen until needed. Careful thawing of the cells is also necessary to maintain cell function. Conditions for collecting samples of whole blood, isolating PBMCs, cryopreservation, and thawing the cells for use in the validation experiments described in this chapter are described in Chapter 10.

IL-5 ELISPOT protocol

Chapter 10 discusses the basic steps of the ELISPOT assay, the assay protocol, results output from the ELISPOT assay and data management tools, data analysis, and result interpretation, which are discussed in detail in Chapter 10. Although these topics are discussed in the context of the IFN-γ ELISPOT assay, they apply to the IL-5 ELISPOT assay with the following exceptions. The optimal number of PBMCs for the IL-5 ELISPOT assay is 300 000 cells per well for stimulation with the antigens, PHA-P, phorbol 12-myristate 13-acetate (PMA) + ionomycin, or superantigen and the plates are incubated for 44 hours. Color development was allowed to proceed for no longer than 15 minutes. Plates were scanned within 2 days and plates can be stored for up to one year at room temperature wrapped in aluminum foil.

11.3 Validation of the IL-5 ELISPOT assay

Similar to the validation of the IFN-γ ELISPOT assay, validation of the IL-5 ELISPOT assay focused on optimizing assay conditions for the specificity of anti-cytokine antibody pairs, cell density, positive control antigen concentration, stability of the profile of cytokine responses with different lots of antigens using fresh and cryopreserved PBMCs, the effects of shipment and storage conditions on the ability of the PBMCs to secrete IL-5 in response to stimulation, stability of the IL-5 cytokine response in PBMCs from normal donors after long term freezing, and assay precision, all using stimulated PBMCs. These validation parameters were used for evaluating several polyclonal agents such as phytohemagglutinin (PHA-P), phorbol 12-myristate 13-acetate (PMA) + ionomycin, and the bacterial superantigens *Staphylococcus aureus* A (SEA) or B (SEB). Validation experiments using PHA-P are described in this chapter

Reagent validation and specificity

The specificity of the ELISPOT assay for the detection of IL-5 secretion was validated in experiments using microtiter wells coated with the IL-5 capture antibody and either the anti-IL-5 or the anti-IFN-γ biotinylated detection antibody. PBMCs from a normal donor were plated into microtiter plate wells coated with the anti-IL-5 capture antibody and cultured for 44 hours with and without PHA-P or PMA + ionomycin. IL-5 spots were observed in stimulated wells when the anti-IL-5 detection antibody was used (data not shown). When the anti-IFN-γ detection antibody was used in place of anti-IL-5 detection antibody, IL-5 secretion could not be detected. IL-5 secretion could not be detected as well in control wells that did not receive activating agent. These results indicate that the IL-5 ELISPOT assay is specific for IL-5 because spots could not be detected when unmatched pairs of anti-cytokine antibodies were used for capture and detection and there is no cross-reactivity with the anti-IFN-γ detection antibody.

Determination of optimal PHA-P concentration

Freshly isolated PBMCs from two normal donor blood samples were treated with concentrations of PHA-P ranging from 1 to 5 μg/mL. PHA-P concentrations of 1 μg/mL results in lower numbers of spots in the two donors whereas 5 μg/mL resulted in <100 spots/well in all donors. The higher concentration of PHA-P can be used to detect both high and low responders to PHA-P. High responders might reach the higher limits of quantification of >600 spots at the 10 μg/mL PHA-P concentration. About 100 spots/well is considered to be a significant IL-5 response and an optimal response reflects sufficient signal to noise ratio for to distinguish low to high responders (Figure 11.1).

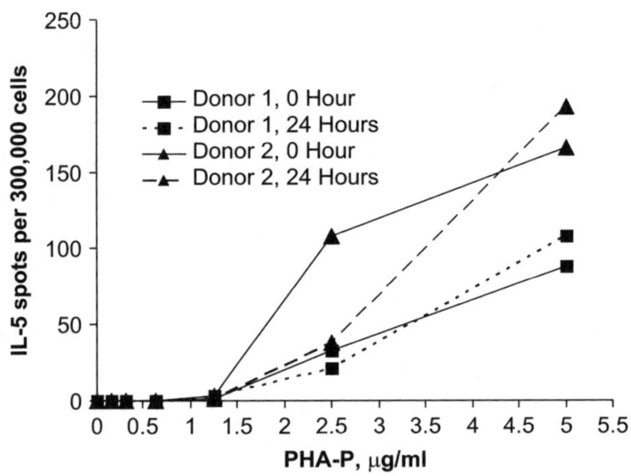

Figure 11.1 Detemination of the optimal PHA-P concentration for IL-5 induction using freshly isolated PBMCs from normal donors. Results shown are representative results from 1 of 3 donors. PHA-P indicates phytohemagglutinin-P; PBMC, peripheral blood mononuclear cell

Determination of optimal cell density and culture time

The optimal cell density for the PHA-P induced IL-5 response by PBMCs isolated from a normal donor was determined by plating varying numbers of cells from 50 000 to 300 000 cells in triplicate wells. The cells were stimulated with 5 μg/mL PHA-P for 20 or 44 hours. The results in Figure 11.2 show that 100 000–300 000 cells per well result in an optimal number of IL-5 spots when the PBMCs are cultured for 44 hours. The IL-5 spots were faint and reduced in number with 20 hours of culture time, compared to the vendor recommended period of 44 hours. In contrast, the spots in the IFN-γ ELISPOT assay become diffuse and darker with longer culture time (see Chapter 11) suggesting different secretion kinetics for IFN-γ and IL-5 cytokines.

Determination of optimal cell density for cryopreserved PBMCs

Cryopreservation of PBMCs facilitates testing of clinical samples in a single experiment to minimize inter-assay variation. To determine the effect of cryopreservation on the IL-5 response of PBMCs to PHA-P stimulation, fresh PBMCs and PBMCs cryopreserved on the same day after on-site blood collection were compared. The fresh and cryopreserved PBMCs were obtained from the same normal donor. The IL-5 responses in both fresh and cryopreserved PBMCs were comparable over the range of cell numbers that were tested (50 000 to 300 000) (data not shown) suggesting that PHA-P induced IL-5 secretion is minimally affected by cryopreservation.

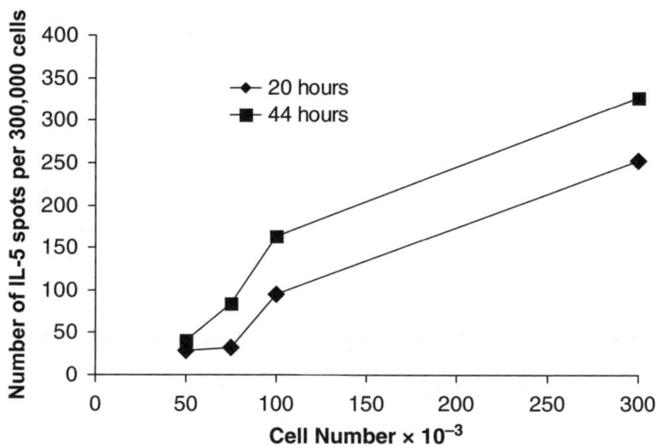

Figure 11.2 Determination of the optimal length of the assay for the PHA-P-induced IL-5 response using PBMCs isolated from a normal donor. PHA-P indicates phytohemagglutinin-P; PBMC, peripheral blood mononuclear cell

Determination of optimal cell density for PBMCs isolated from overnight shipped blood samples

PBMCs isolated from overnight shipped samples gave an optimal response to PHA-P when 300 000 cells were used (Table 11.1). The density of 400 000 cells cannot be used for the IL-5 ELISPOT assay because 500 IL-5 spots were produced by PBMCs from one donor at this high cell number reaching the upper limit of quanitification. Spots numbering >600 cannot be accurately counted (see the section on Limit of quantification) in the ELISPOT assay.

These findings suggest that the PBMCs can be isolated and cryopreserved when clinical study samples are received from the clinical study sites and then tested at a later date to conduct an IL-5 ELISPOT assay in a batch mode.

Comparison of IL-5 responses from PBMCs cryopreserved from same day vs. day 1 or day 2 received blood samples

To simulate the handling conditions to which clinical study samples will be subjected, PBMCs were isolated from whole blood samples from 10 normal donors and cryopreserved before and after overnight shipment of the blood at ambient temperature. The IL-5 response to PHA-P in PBMCs isolated and cryopreserved from overnight-shipped blood samples were reduced compared with the response of cells cryopreserved on the day of sample collection (Table 11.2). This response was further reduced in samples received on the second day (Table 11.2 and shown for one donor in Figure 11.3). The reduced response of shipped samples cannot be directly attributed to slightly reduced viability at 48 hours compared with PBMCs from freshly drawn blood (Table 11.3) because only viable cells from all samples were included in the assay. Therefore, the reduced functionality of PBMCs, attributable to lower number and functionality of antigen presenting cells preserved after shipment and freezing, is the likely reason for the responses observed with shipped samples (Kumar and Satchidanandam, 2000). Based on these results samples received on the second day after collection are excluded from analysis.

Stability of the IL-5 response in long-term cryopreserved samples

The stored or shipped samples must be part of the same blood draw and must be tested over time using the same reagents and protocol. The duration of the stability studies should encompass the duration of the planned clinical study in which the assay will be used.

The IL-5 response from PBMCs was compared with the response when the cells were freshly cryopreserved and after 14 months of storage. The

Table 11.1 Effect of cell density on IL-5 secretion by cryopreserved PBMCs from overnight-shipped samples stimulated with PHA-P[a,b]

Cell number	Donor 1	Donor 2	Donor 3	Donor 4	Donor 5	Donor 6	Donor 7	Donor 8	Donor 12	Donor 13	Donor 14	Donor 15	Donor 16	Donor 17	Donor 18
100 000				85								54	64	85	
200 000	120	124	272	60	147	120	124	272	124	272	120	80	158	60	147
300 000	102	309	290	92	244	102	309	290	309	290	102	116	194	92	244
400 000	180	321	500	92		180	321	500	321	500	180	130	292	92	

[a] Values are the number of IL-5 spots per the number of PBMCs.
[b] PBMC indicates peripheral blood mononuclear cell; PHA, phytohemagglutinin-P.

Table 11.2 PHA-P induced IL-5 secretion by cryopreserved PBMCs from shipped samples[a]

Donor	Before shipment	After sample receipt	
		Day 1	Day 2
Donor 1	258	151	54
%Change (before and after)		41.4	79.04
Donor 2	386	242	55
%Change (before and after)		37.39	85.75
Donor 3	213	92	31
%Change (before and after)		56.81	85.45
Donor 4	34	30	1
%Change (before and after)		11.76	97.06
Donor 5	140	36	8
%Change (before and after)		74.22	94.51
Donor 6	62	42	3
%Change (before and after)		32.43	95.68
Donor 7	62	79	15
%Change (before and after)		−27.91	76.42
Donor 8	29	17	0
%Change (before and after)		43.10	98.85
Donor 9	12	18	1
%Change (before and after)		−50.00	91.67
Donor 10	34	30	1
%Change (before and after)		11.76	97.06

[a]PHA-P indicates phytohemagglutinin-P; PBMC indicates peripheral blood mononuclear cell.

Table 11.3 Cell viability in cryopreserved PBMCs from shipped samples[a]

Donor	Fresh	24 hour	48 hour
1	97.90	94.50	91.20
2	92.70	88.40	84.50
3	86.20	74.20	82.50
4	94.20	83.20	95.60
5	94.50	89.10	89.60
6	94.80	91.70	86.90
7	90.00	77.70	83.70
8	94.20	83.20	95.60
Mean	93.06	85.25	88.70
% change (before and after)		8.39	4.69

[a]PBMC indicates peripheral blood mononuclear cell.

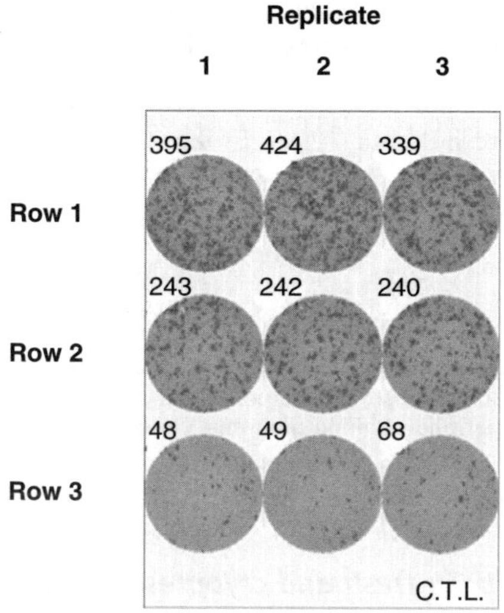

Figure 11.3 The IL-5 response of cryopreserved PBMCs from a normal donor stimulated with PHA-P. Each row of wells is a sample that has been plated in triplicate. Row 1: PBMCs isolated and cryopreserved on the day the sample was collected. Row 2: PBMCs isolated and cryopreserved after overnight shipment of the whole blood sample. Row 3: PBMCs isolated and cryopreserved from whole blood samples that were in shipment for two days. PBMC indicates peripheral blood mononuclear cell; PHA-P, phytohemagglutinin-P.

Table 11.4 Effect of long-term cryopreservation on PHA-P induced IL-5 response by PBMCs frozen on the day of sample collection[a,b]

Donor	Months		Mean	SD	%CV
	0	14			
42318	200	283	242	8.49	3.51
36019	188	239	214	36.06	16.89

[a]Values are the number of IL-5 spots per 300 000 PBMCs.
[b]PHA-P indicates phytohemagglutinin-P; PBMC, peripheral blood mononuclear cell; %CV, percent coefficient of variation.

results suggest that the IL-5 response can be induced in these cryopreserved PBMCs stored for more than 1 year (Table 11.4). Thus, PBMCs from a normal donor can be used as a positive control in all ELISPOT assays conducted for a particular clinical study. Though this use of normal control confirms that the assay was performed accurately, it does not provide any information on the integrity of cryopreserved clinical samples being evaluated. Viability assessments do provide this information to some extent.

Table 11.5 Intersubject variability of the IL-5 response in fresh PBMCs and in cryopreserved PBMCs from overnight-shipped samples[a,b]

	Fresh 0 hour ($n = 10$)	Frozen 0 hour ($n = 20$)	Frozen 24 hour ($n = 22$)
Range	110–609	50–470	14–414
Mean	249	204	159
SD	155.56	106.03	146.68
%CV	62.36	52.10	92.04

[a]Values are the number of IL-5 spots per 300 000 PBMCs.
[b]PBMC indicates peripheral blood mononuclear cell; %CV, percent coefficient of variation.

Intersubject variability in fresh and cryopreserved PBMCs

PBMCs isolated from freshly collected whole blood and cryopreserved PBMCs isolated and cryopreserved on the day of sample collection or following overnight shipment were stimulated with PHA-P and IL-5 secretion was measured. There was considerable variation in the IL-5 response of the PBMCs to PHA-P stimulation. The %CV ranged from 52.1 to 92.04 for fresh PBMCs, freshly cryopreserved PBMCs, and PBMCs isolated and cryopreserved after overnight shipment of whole blood samples (Table 11.5). Because of the high intersubject variability of the IL-5 response, the IL-5 secretion profile of each patient will be compared to his/her own baseline sample in clinical studies (pre-treatment vs. post-treatment response).

Assay precision in the IL-5 ELISPOT assay with PHA-P stimulation

Intra-assay variability

Cryopreserved PBMCs isolated from a same day shipped sample were assayed in parallel in triplicate wells on three different plates. PHA-P stimulated IL-5 secretion between the triplicate wells in all three plates resulted in SDs ranging from 9.29 to 35.12 and %CVs ranging from 1.89 to 14.03 (Table 11.6). In addition, cryopreserved PBMCs from four overnight shipped normal samples were assayed in triplicate wells to evaluate the intraassay variability of IL-5 ELISPOT detection. The determination of IL-5 ELISPOTs in triplicate wells was precise as indicated by the pooled SD of 16.26 and %CV of 11.46 for 11 donors (Table 11.7). These results indicate that variability within this assay is acceptably low and justifies the analysis of clinical samples in duplicate in the event of limited sample

Table 11.6 Intra-assay variability in the IL-5 ELISPOT assay[a,b]

Plates	Replicate 1	Replicate 2	Replicate 3	Mean	SD	%CV
Plate 1	498	481	496	492	9.29	1.89
Plate 2	510	466	502	493	23.44	4.76
Plate 3	450	452	484	462	19.08	4.13
Plate 1	191	193	232	205	23.12	11.26
Plate 2	287	217	247	250	35.12	14.03
Plate 3	272	224	282	259	31.01	11.96
Plate 1	190	225	228	214	21.13	9.86
Plate 2	239	249	289	259	26.46	10.22
Plate 3	251	233	267	250	17.01	6.79
Plate 1	413	348	381	381	32.50	8.54
Plate 2	351	388	371	370	18.52	5.01
Plate 3	378	361	395	378	17.00	4.50

[a]Values are the number of IL-5 spots.
[b]PBMC indicates peripheral blood mononuclear cell.

Table 11.7 Intra-assay variability in the IL-5 ELISPOT assay[a,b]

Donor	Replicate 1	Replicate 2	Replicate 3	Mean	SD	%CV
1594	285	335	306	309	25.11	8.13
1596	283	291	295	290	6.11	2.11
1878	119	92	85	99	17.95	18.20
1415	100	117	90	102	13.65	13.34
1806	391	377	438	402	31.95	7.95
1825	426	356	415	399	37.64	9.43
1605	22	31	22	25	5.20	20.78
1415	88	72	85	82	8.50	10.41
1796	51	63	45	53	9.17	17.29
1414	367	369	389	375	12.17	3.24
1596	88	70	67	75	11.36	15.14
Pooled					16.26	11.46

*Values are the number of IL-5 spots per 300 000 PBMCs.
†PBMC indicates peripheral blood mononuclear cell.

availability because of a low cell count or poor recovery of the cells after thawing.

Inter-assay precision

The IL-5 response of cryopreserved PBMCs to PHA-P stimulation was studied on the same day using two different plates to assess interassay (interplate) variability. Cells were isolated using whole blood samples from four normal donors that were shipped the next day to mimic the timeframe of clinical sample shipment and delivery. A pooled SD and %CV of 9.37 and 19.82, respectively, (data not shown) were obtained, which indicate that the IL-5 ELISPOT assay has low interplate variability.

Inter-assay variability was also assessed in assays performed on three consecutive days using cryopreserved PBMCs from overnight shipped samples as well as after long term storage to evaluate the possibility of using the cryopreserved PBMCs from normal donors as positive controls for testing clinical samples. Because different cryovials of PBMCs from the same lot of cells was used for these studies, the within day or inter-day results can be considered inter-assay variability. The pooled inter-assay %CV for the IL-5 response from cryopreserved PBMCs isolated from overnight shipped samples was 23.28 (Table 11.8). The use of a different cryovial of PBMCs on each assay day contributed to part of this interday or interassay variability. Even though the cryovials were prepared from the same lot of PBMCs, sample to sample variability does exist in cell viability and function.

Inter-operator precision

Two trained operators evaluated the IL-5 responses to PHA-P in cryopreserved PBMCs from seven normal donors isolated from samples that had been shipped overnight. The SD ranged from 0.71 to 28.99 and the %CV values ranged from 2.15 to 28.7 for the IL-5 response to PHA-P stimulation for the two operators (Table 11.9).

Table 11.8 Inter-assay (inter-day) variability of the IL-5 ELISPOT assay in PBMCs cryopreserved from overnight-shipped samples[a,b]

Donor	Day 1	Day 2	Day 3	Mean	SD	%CV
1415	47	88	49	61	23.12	37.69
1596	68	75	94	79	13.45	17.03
1843	59	68	57	61	5.86	9.55
42318	282	283	300	288	10.22	3.55
36019	346	239	274	286	54.19	18.92
35265	97	59	235	130	92.89	71.30
46200	458	415	439	437	21.63	4.95
Pooled					31.62	23.28

[a]Values are the number of IL-5 spots per 300 000 PBMCs.
[b]PBMC indicates peripheral blood mononuclear cell.

Table 11.9 Inter-operator variability of the IL-5 ELISPOT assay[a]

Donor	Operator 1	Operator 2	Mean	SD	%CV
	Mean				
1	33	42	38	6.6	17.52
2	63	87	75	16.5	22
3	33	32	33	0.71	2.15
4	28	21	25	5.42	22.13
5	78	110	94	22.16	23.57
6	81	122	101	28.99	28.70
7	125	156	140	21.92	15.64

[a]PBMC indicates peripheral blood mononuclear cell.

Effect of different lots of reagents and assay components on PHA-P induced IL-5 responses

Effect of different lots of PHA-P

Cryopreserved PBMCS from four normal donor samples were plated into triplicate wells (300 000 cells/well) and stimulated with 5 μg/mL PHA-P from two different lots. The IL-5 response from both lots was comparable in all four donors (data not shown). Therefore, new lots of PHA-P can be used during a clinical study without re-titrating the mitogen concentration or cell density. However, all samples from one subject collected at different time points should be tested with the same lot of mitogen.

Effect of different lots of assay plates

Cryopreserved PBMCs from four overnight shipped samples were plated into triplicate wells (300 000 cells/well) of plates from two different lots of microtiter plates. The cells were stimulated with PHA-P and the IL-5 response was measured. Comparable results were obtained in both assay plates for all four donors (data not shown). Cells incubated without PHA-P gave a negligible background response that was similar in both assay plates as well. Therefore, different lots of assay plates can be used during a clinical study without re-titrating the mitogen concentration or cell density. However, it is recommended that all samples from one subject collected at different time points should be tested using the same lot of assay plates to minimize or exclude any variability that may be contributed by factors other than the variability of samples.

The more recently available high-binding ELISPOT plates were compared with regular binding ELISPOT plates for use in the IL-5 ELISPOT assay using PBMCs from two donors. In this experiment, the

Table 11.10 Comparison of regular and high-binding ELISPOT plates in the IL-5 ELISPOT assay

Donor	Regular Millipore			High-binding Millipore		
	24 hour	48 hour	1 week	24 hour	48 hour	1 week
1	207	229	181	468	489	439
2	152	147	170	272	293	277

stability of plates coated with anticytokine antibody for up to 1 week before use in the assay was assessed as well. The performance of the regular and high-binding plates coated with anticytokine antibody for up to 1 week before use in the IL-5 ELISPOT assay was similar to plates coated 24 hours before use (Table 11.10). About twice the number IL-5 spots appeared on the high-binding plates compared with the regular-binding plates for all coating conditions. Thus, plates can be coated for up to 1 week prior to the IL-5 ELISPOT assay and the plates will provide good results. High-binding plates perform better in the assay than regular-binding plates.

Effect of assay components from different sources

IL-5 ELISPOT assay components, such as the IL-5 antibody pairs (capture and detection antibodies) and streptavidin HRP are routinely purchased as individual items from BD Biosciences-Pharmingen, San Diego, CA, and as well as assay plates from Millipore Corporation, Billerica, MA. On occasion, it is necessary to use the human IL-5 ELISPOT set from BD Biosciences-Pharmingen because the antibody pairs or the assay plates are not available from the vendors.

To compare assay components from different sources, cryopreserved PBMCs isolated from two normal donor samples that had been shipped overnight were plated in triplicate into assay plates from two different sources (MultiScreenTM–IP, Millipore Corporation, Billerica, MA, and BD ELISPOT set plate, San Diego, CA) and the PHA-P induced IL-5 response was measured using individual components or reagents from the human IL-5 ELISPOT set. Comparable results were obtained with both the individual assay components and the BD ELISPOT set materials. The %CVs ranged from 8.70 to 14.30. Therefore IL-5 ELISPOT assays can be performed without re-titrating the mitogen concentration or cell density when reagents purchased as individual components or as a set are used. The use of individual components is cost effective and is recommended for assaying clinical samples with the caveat that all samples collected during the course of the clinical study from each individual subject should be tested using reagents from one source.

Stability of capture antibody-coated plates

To determine the reproducibility of the IL-5 response of PBMCs to stimulation with PHA-P in plates that have been coated with anticytokine antibody and stored at 4 °C for different lengths of time, cryopreserved PBMCs isolated from a normal donor sample after overnight shipment were added to assay plates. The plates had been coated with anticytokine antibody 1 or 2 days before the study. The IL-5 response was reduced in wells that had been coated 2 days before the assay. These findings would be of concern with samples that have a low IL-5 response to PHA-P stimulation are tested. Therefore, only plates coated less than 1 day before the assay date can be used for the assay to accurately determination of IL-5 response in all samples.

11.4 Limit of quantification and sample and data acceptance criteria

Limit of quantification

The lower limit of quantification (LLOQ) was determined to be a mean spot number of 10 based on the results observed in different assays. Representative results from several of these assays are shown in Table 11.7 and Figure 11.4(a). It was observed that %CV tended to be higher when lower spot numbers occurred in triplicate wells and there seemed to be less variability when spot numbers were greater than 20. To be able to capture low frequency responses that occur mainly to a specific antigen, the LLOQ has been set at mean spot number of 10.

Based on the number and morphology of the spots, the upper limit of quantification (ULOQ) between triplicate wells is 600 spots/well (Figure 11.4(b)). This number of spots has a %CV $\leq$30. The responses resulting in spot numbers above 600 are usually an estimate based on mean spot sizes and are not accurate.

Sample and data acceptance criteria

Assay conditions

The assay conditions for the IL-5 ELISPOT assay are the same as the assay conditions described in Chapter 10 for the IFN-γ ELISPOT assay.

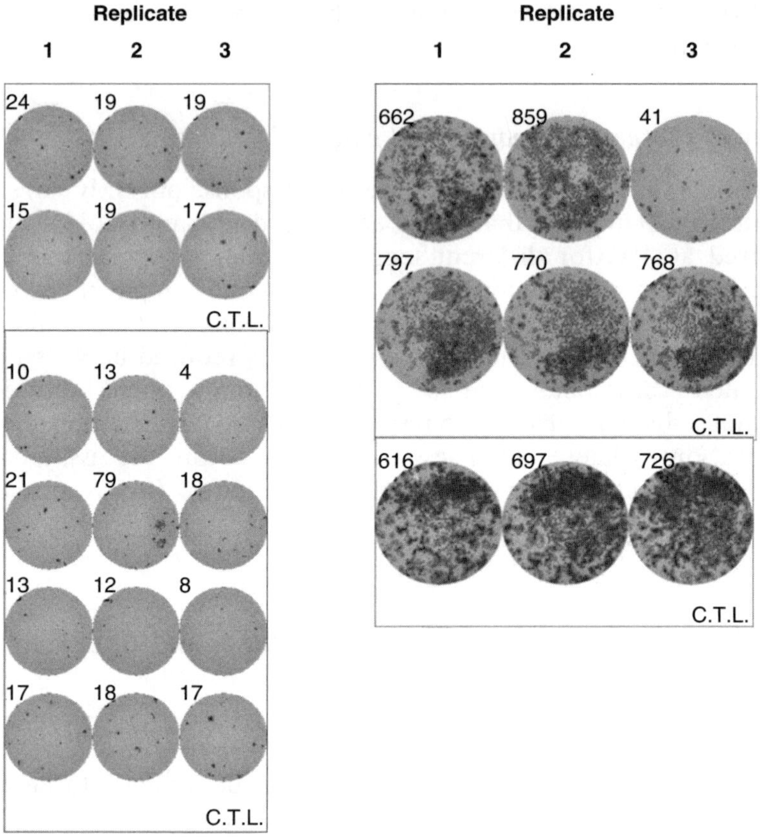

Figure 11.4 Examples of IL-5 ELISPOT wells demonstrating the lower limit of quantification (LLOQ) (left) and the upper limit of quantification (ULOQ) (right). Each row of wells is a sample that has been plated in triplicate. The LLOQ was set at a mean of 10 spots/well because lower spot numbers cannot readily be distinguished. The ULOQ was set a mean of 600 spots per well. Spot morphology and the confluence of the spots at higher spot numbers reduce the accuracy of the spot count.

Clinical sample and data acceptance criteria

The clinical sample and data acceptance criteria for the IL-5 ELISPOT assay are the same as the criteria described in Chapter 10 for the IFN-γ ELISPOT assay.

Limit of quantification and acceptance criteria

- The LLOQ was determined to be a mean spot number per well of 10.
- Based on the number and morphology of the spots, the ULOQ with assay precision of $\leq$30% CV between triplicate wells is 600 spots/well.

Table 11.11 Determination of LLOQ[a,b]

Replicates			Mean	SD	%CV
1	2	3			
24	19	19	21	2.89	13.97
15	19	17	17	2.00	11.76
10	13	4	9	4.58	50.92
21	79	18	39	34.39	87.42
13	12	8	11	2.65	24.05
17	18	17	17	0.58	3.33

[a]Values are the number of IL-5 spots per 300 000 PBMCs.
[b]LLOQ indicates lower limit of quantification; PBMC, peripheral blood mononuclear cell; %CV, percent coefficient of variation.

Table 11.12 Lower limit and upper limit of quantification for the IL-5 ELISPOT assay

	Stimulation	Cells per well	SFC per well[a]	SFC/PMC
LLOQ	Unstimulated	300 000	<10	<33
	PHA-P	300 000	<10	<33
	PMA + ionomycin	300 000	<10	<33
	SEB	300 000	<10	<33
ULOQ	Unstimulated	300 000	>600	>2000
	PHA-P	300 000	>600	>2000
	PMA + ionomycin	300 000	>600	>2000
	SEB	300 000	>600	>2000

[a]SFC indicates spot forming cells; PMC, per million cells; LLOQ, lower limit of quantification; ULOQ, upper limit of quantification; PHA-P, phytohemagglutinin-P; PMA, phorbol-12-myristate 13-acetate; SEB, staphylococcal enterotoxin B.

- Since the number of spot forming cells (SFC) per well is represented as SFC/PMC (per million cells) for clinical study samples, the SFC of <10 (LLOQ) and >600 (ULOQ) are represented in the SFC/PMC table of the data analysis template based on the cell number plated as shown in Tables 11.11 and 11.12.

11.5 Summarizing remarks

An ELISPOT assay for determining the number of IL-5 spots or IL-5 SFC using PBMCs isolated and cryopreserved from human peripheral whole blood has been validated for use with clinical samples received the same day or next day after collection at clinical study sites. The accuracy and specificity of IL-5 spot detection was demonstrated by comparing the results using a detection antibody for IL-5 compared with a detection antibody for IFN-γ. The IL-5 response in PBMCs isolated the same day or the day after collection was not affected by cryopreservation for periods of time up to one year. Assay conditions were optimized for assay length and cell density of cryopreserved PBMCs isolated from samples received the same day and the day after collection of the whole blood sample. Cryopreserved PBMCs from overnight-shipped samples secreted IL-5 with high reproducibility in triplicate wells with low intra assay variability. Although interday (interassay) variability was higher than intraassay variability, it should not impact the accuracy of the results if all samples collected at all time points from a patient during a clinical study are analyzed on the same day. Likewise, interplate variability was determined to be minimal suggesting that multiple plates can be used if needed to test all of a subject's samples on the same day without adversely affecting the results. Interoperator variability was also low confirming the precision of the assay procedure. The final results from patient samples are expressed as mean SFC per million PBMCs and as percent change from the baseline or pre-treatment samples. The IL-5 ELISPOT assay can be used to determine the number of IL-5 secreting cells in PBMCs isolated from whole blood samples from clinical study patient samples, collected off-site and shipped to the testing lab to arrive the same day or the following day after collection. The cells must be cryopreserved immediately upon receipt.

References

Bennouna J, Hildesheim A, Chikamatsu K, Gooding W, Storkus WJ and Whiteside TL (2002). Application of IL-5 ELISPOT assays to quantification of antigen-specific T helper responses. *J Immunol Methods*, **261**, 145–156.

Clay TM, Hobeika AC, Mosca PJ, Lyerly HK and Morse MA (2001). Assays for monitoring cellular immune responses to active immunotherapy of cancer. *Clin Cancer Res*, **7**, 1127–1135.

Correale J and Fiol M (2004). Activation of humoral immunity and eosinophils in neuromyelitis optica. *Neurology*, **63**, 2363–2370.

Genain CP, Abel K, Belmar N, Villinger F, Rosenberg DP, Linington C, Raine CS and Hauser SL (1996). Late complications of immune deviation therapy in a nonhuman primate. *Science*, **274**, 2054–2057.

Jansson A, Ernerudh J, Kvarnstrom M, Ekerfelt C and Vrethem M (2003). Elispot assay detection of cytokine secretion in multiple sclerosis patients treated with interferon-beta1a or glatiramer acetate compared with untreated patients. *Mult Scler*, **9**, 440–445.

Kanik KS, Hagiwara E, Yarboro CH, Schumacher HR, Wilder RL and Klinman DM (1998). Distinct patterns of cytokine secretion characterize new onset synovitis versus chronic rheumatoid arthritis. *Rheumatol* , **25**, 16–22.

Kreher CR, Dittrich MT, Guerkov R, Boehm BO and Tary-Lehmann M (2003). CD4+ and CD8+ cells in cryopreserved human PBMC maintain full functionality in cytokine ELISPOT assays. *Immunol Methods*, **278**, 79–93.

Kumar P and Satchidanandam V (2000) Ethyleneglycol-bis-(beta-aminoethylether) tetraacetate as a blood anticoagulant: preservation of antigen-presenting cell function and antigen-specific proliferative response of peripheral blood mononuclear cells from stored blood. *Clin Diagn Lab Immunol*, **7**, 578–583.

Kvarnstrom M, Jenmalm MC and Ekerfelt C (2004). Effect of cryopreservation on expression of Th1 and Th2 cytokines in blood mononuclear cells from patients with different cytokine profiles, analysed with three common assays: an overall decrease of interleukin-4. *Cryobiology*, **49**, 157–168.

Okamoto Y, Gotoh Y, Shiraishi H and Nishida M (2004). A human dual-color enzyme-linked immunospot assay for simultaneous detection of interleukin 2- and interleukin 4-secreting cells, *Int Immunopharmacol*, **4**, 149–156.

Pelfrey CM, Rudick RA, Cotleur AC, Lee JC, Tary-Lehmann M and Lehmann PV (2000). Quantification of self-recognition in multiple sclerosis by single-cell analysis of cytokine production, *J Immunol*, **165**, 1641–1651.

Ronnelid J, Berg L, Rogberg S, Nilsson A, Albertsson K and Klareskog L (1998). Production of T-cell cytokines at the single-cell level in patients with inflammatory arthritides: enhanced activity in synovial fluid compared to blood, *Br J Rheumatol*, **37**, 7–14.

12 Validation of the Cylex technology to measure T and B cell activation capacity in clinical trials

Marielena Mata*, Thomas Lohr and Jaymala Patel
Centocor Research and Development, Inc., Radnor, PA 19087, USA

12.1 Introduction

Lymphocytes play a critical role in the function of the immune system. They can be involved in aspects of both innate and adaptive immunity and have been shown to play a role in defense against infection through the production of antibodies, and coordination of immune signaling pathways. Lymphocytes can be further classified into T and B-lymphocytes based on function as well as surface marker expression. B-lymphocytes comprise

* Author to whom correspondence should be addressed.

Validation of Cell-Based Assays in the GLP Setting: A Practical Guide Editors, Uma Prabhakar and Marian Kelley

approximately 5–15% of all circulating lymphocytes while most circulating lymphocytes (80%) are T cells. Exposure of lymphocytes to stimulus results in activation and expansion of the population. Stimulated lymphocytes first undergo an influx of ions and increased adenosine triphosphate (ATP) synthesis followed by surface receptor clustering, RNA synthesis, cytokine production and release and DNA replication (Paul, 2003).

During different disease states or under different drug treatments, functionality of the lymphocyte compartment can be compromised intentionally or unintentionally. In the case of transplantation patients, suppression of T cell lymphocyte function is required to prevent transplant rejection. Alternatively, immunosuppression can be the unintended outcome of certain oncology chemotherapy treatments. Additionally, lymphocyte function monitoring is valuable during development of new drugs as possible markers of efficacy and/or safety.

A number of assays and methodologies exist to evaluate immune function such as flow cytometry, ELISPOT, proliferation assays, each with its own limitations (Keilholz et al, 2002). Flow cytometry methods are valuable for cell enumeration but are not always able to assess the functional capacity of such cells. From a functional point of view, proliferation assays and ELISPOTs provide the best readout for the ability of lymphocytes to respond to a given antigen but they are cumbersome and difficult to validate (Keilholz et al, 2002). Another approach is to look at early activation signals such as ATP release (Sottong et al, 2000).

The Cylex assay takes advantage of these early steps during the activation process to evaluate functional activity of lymphocytes to antigens and mitogens. It measures the early activation response by detecting intracellular ATP synthesis of specific lymphocyte populations selected from blood by monoclonal antibody coated magnetic beads. The amount of ATP present in stimulated blood specimens is a measure of lymphocyte activity. An additional advantage of using a kit is the ability to validate the assay in a regulated environment after the method has been optimized by the vendor. In fact, the Cylex ImmuKnow assay for cluster of differentiation 4 (CD4) activity has been approved by the FDA for the monitoring of the patient's immune system for post-transplantation drug follow-up (Hooper et al, 2005; Knight et al, 2005; Kowalski et al, 2003; Zeevi et al, 2005).

In addition, Cylex has developed other assays to monitor CD3 and CD19 cell activity using the same platform technology. Such assays could be implemented to monitor the immune activity in AIDS and cancer patients, and those with infectious diseases, autoimmune or other immune system disorders. With that purpose, we validated two assays, the CD4 and CD19 assays to monitor the immune capacity of patients in clinical trials.

The FDA provides very clear guidelines for the validation of Bioanalytical assays (FDA, 2001), yet many of these guidelines are not applicable to cell-based assays or do not cover all aspects that need to be evaluated in cell-based assays. In addition, the application of these assays to understand the pharmacodynamics of new compounds provides additional flexibility from a regulatory point of view. Nonetheless, in order for cell-based assays to provide the appropriate value, they need to be developed and validated using more stringent criteria than has been applied up to this point.

In this validation, we try to address some of these issues, including the qualitative nature of the assay, even in the presence of a quantitative component, stability of samples not only in-house but as it may be affected by shipping conditions and the interference of the drug compounds being evaluated in the clinical trials where these assays would be employed. Finally, the validation of each assay should be carefully planned to address all the different variables that need to be evaluated as discussed in other chapters in this book.

12.2 Materials and methods

Sample collection

Whole blood was collected aseptically under informed consent from healthy volunteers within the Centocor in house normal donor bank or purchased from Biological Specialty Corporation (Colmar, PA). Blood was drawn by venipuncture into tubes containing sodium heparin anticoagulant (VWR, Bridgeport, NJ). Each tube was mixed thoroughly by inversion and allowed to sit, tilted on its side, in a rack at room temperature (20–25 °C). Blood received from Biological Specialty Corporation was shipped in sodium heparin tubes either at ambient temperature or with ice packs (4 °C). Whole blood was used for assay set up within 48 hours post collection. Hemolyzed or clotted blood specimens were not used and excessive shaking or mixing (e.g., rocker or rotator) was avoided.

Assay method

Whole blood is incubated overnight either unstimulated or in the presence of the antigens/mitogens phytohemaglutinin (PHA) (Cylex), lipopolysaccharide (LPS) (Sigma, Saint Louis, MI) Staphylococcal Enterotoxin A (SEA) (Sigma) or Staphylococcal Enterotoxin B (SEB) (Sigma). Following incubation, $CD19^+$ B or $CD4^+$ T lymphocytes are collected by positive exclusion of cells bound to appropriately coated magnetic beads. Captured cells are lysed and liberated ATP concentrations are measured in the supernatant of each assay well. ATP concentrations of unknown samples are interpolated from a kit-provided standard calibration curve and reported in ng/mL. Cylex Assay Kits for CD4 and CD19, PHA-L, Magnet Tray,

Eight Channel Aspirator, Microplate Shaker and Data Analysis software (Version 2.1) were obtained from Cylex Incorporated (Columbia, MD).

A step-by-step description of the Cylex assay procedure is provided in the kit. A brief summary is described here.

1. On the first day, dilute each stimulant reagent using the provided Sample Diluent to achieve the desired concentration (PHA, 2.5 µg/mL; LPS, 0.1µg /mL; SEA, 100 ng/mL; SEB, 100 ng/mL).
2. Dilute the required amount of blood 1:4 in Sample Diluent. Add diluted stimulant and diluted blood into assay plate wells according to plate template, mix and incubate for 15–24 hours at 37 °C / 5% CO_2 / 90% relative humidity.
3. The next day, prepare CD19 Dynabeads as described. Resuspend beads and dispense the necessary volume of CD 19 beads into a 1275 tube. Wash bead by placing the 12×75 mm tube into the assembled Magnet Tray and aspirating the fluid from the tube after all the magnetic beads have migrated to the side of the tube then adding 1 mL of Wash Buffer to the tube for a total of three wash cycles.
4. After washing, resuspend bead in appropriate volume of Wash Buffer for the next step. CD4 dynabeads do not require prior preparation.
5. Remove the assay plate from the incubator, resuspend the cells using a plate shaker and add appropriate volume of dynabeads. Incubate cells with dynabeads at room temperature (18–25 °C) for 30 minutes with intermittent shaking.
6. Place individual well strips of the assay plate on the assembled magnet tray to allow for positive selected cells bound to the magnetic beads to collect on the side of the wells.
7. Aspirate the unbound cells in solution from each well using an eight-channel aspiration manifold taking and wash three times to eliminate unbound cells and interfering substances from the wells.
8. After washing, add Lysis Reagent into each well, shake and replace in the magnet tray.
9. Transfer 50 µL/well of cell sample and calibration curve panel sample to the corresponding wells of the Measurement Plate taking care not to disturb or remove the magnetic beads from the side of the well.
10. Add 50 µL/well of each concentration of the Calibration Curve Panel into the Measurement Plate according to the plate template. Add 150 µL of Luminescence Reagent into each well of the Measurement Plate and incubate for 3 minutes at room temperature.
11. Following this incubation read the plate on the luminometer (Top Count), (PerkinElmer, Shelton, CT) within 10 minutes.
12. Calculate the amount of ATP present in the cell lysate by interpolating from an ATP Calibration Curve generated for each Measurement

Plate. Relative Light Units (RLU) for samples are obtained from the luminometer and converted to ng/mL ATP based on comparable RLU signals in the calibrators.

13. The RLU data from the luminometer is converted and analyzed using the in vitro Cell-Mediated Immunity Data Analysis software provided by Cylex. Briefly, ATP Calibrator Concentrations vs. RLU values are plotted on a log–log scale, and linear regression analysis generates a calibration curve from which unknown sample values can be calculated.

Data analysis and documentation

GraphPad Prism software was used for statistical data analysis. Data analyzed by GraphPad prism software included incubation time point data, blood storage at room temperature and blood shipped from Biological Specialty Corporation.

All data generated from the validation of these assays was properly documented in notebooks and supplemental binders that were archived. Raw data was checked against the validation report to ensure accuracy. All steps of the assay validation were conducted following corporate and departmental Standard Operating Procedures.

12.3 Results and discussion

Standard curve

The manufacturer's recommended standard curve range for both assays is from 0 to 1000 ng/mL. Values for ATP concentration are reported as ng/mL based on the calibration curve plot of log RLU versus log ATP (ng/mL).

Stability

Incubation time points

Optimal cell activation time was evaluated by stimulating whole blood samples from normal donors with PHA for 15, 18, 21 or 24 hours. A positive response was elicited after 15 hours of stimulation and the response remained stable, without significant change (CD19, $p = 0.1001$; CD4 $p = 0.0969$) for up to 24 hours of stimulation although the actual levels of ATP measured at different times were variable (Figure 12.1). Therefore, for comparison purposes, it is important to assess sample sets under similar conditions.

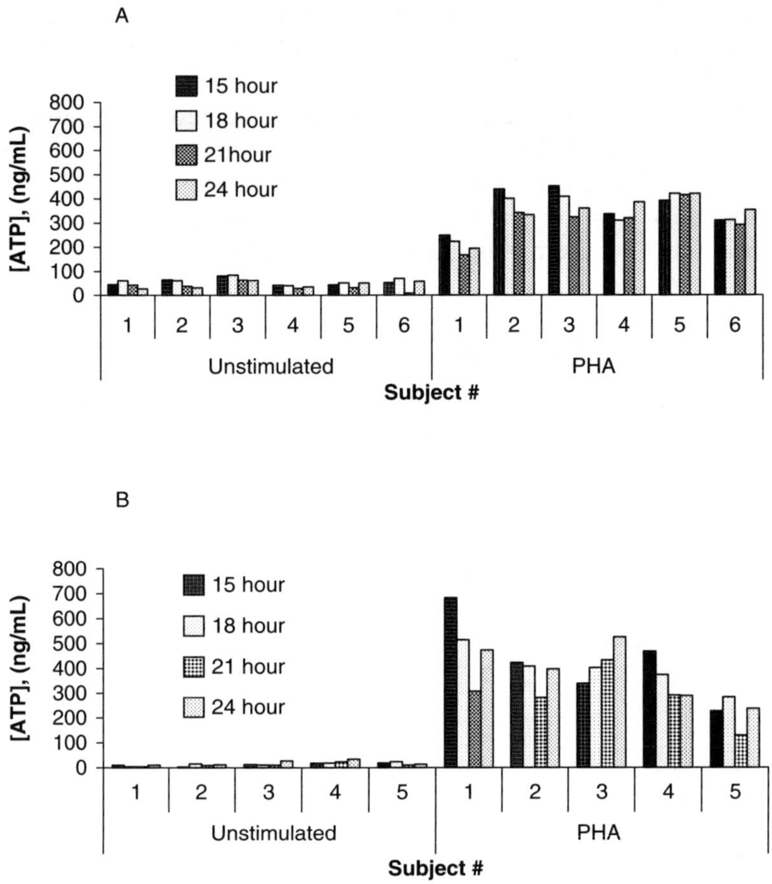

Figure 12.1 Optimal stimulation time. Samples were stimulated with PHA (2.5ug/mL) for 15,18, 21 or 24 hours and analyzed using the appropriate Cylex kit (*A*, Cylex CD19 kit, *B*, Cylex CD4 kit).

Stability of samples at room temperature

Whole blood sample stability was evaluated at room temperature in five whole blood samples from healthy volunteers. Samples were analyzed following 24 hours and 48 hours of storage at room temperature. Samples stored up to 48 hours post draw at room temperature continue to maintain the activation response although the actual levels of ATP measured at different times were variable. In the particular case of CD19, some donor samples did exhibit a decrease in response after 48 hours of storage, however this change was found not to be significant ($p = 0.25$).

Shipment stability

Sample shipping stability was assessed using whole blood samples from five to six healthy donors shipped from an outside vendor at either ambient temperature and on icepacks (~4 °C). All samples were shipped for next day delivery in insulated Styrofoam boxes. Samples shipped at ambient temperatures showed similar stimulation responses at 24 and 48 hours post draw, without significant change (CD19+, $p = 0.0594$; CD4+, $p = 0.3457$) although the actual levels of ATP measured at different times were variable (Figure 12.2(A, B)). Samples shipped on ice did not maintain response to stimulation past same day of collection. It is therefore recommended that clinical samples be shipped at ambient temperature and evaluated in this assay within 48 hours of being drawn (Figure 12.2(C, D)).

Diluted stimulant stability

Since the stimulant reagent PHA used to stimulate the CD19+ B cells needs to be diluted prior to the assay, diluted stimulant stability was determined by evaluating stimulant performance in the assay following dilution and storage at 4 °C for 1, 3, and 5 days prior to assay set up. PHA was diluted and placed at 4 °C for the designated length of time. No significant changes in assay performance were recorded with stimulant stored at 4 °C for up to 5 days. The PHA provided in the CD4 kit does not require dilution and was not evaluated in that validation.

Assay plate stability and storage

Stability of the completed assay plate was evaluated for potential long-term storage at –20 °C. Assay plates with five to six samples from healthy human donors were evaluated on days 0, 10, 20, and 30. Two-experiment plates were required to provide enough sample volume to be read at the different time points. Minimal variability was observed at days 10, 20 and 30 when compared to data from day 0 from their respective plate. Percent coefficients of variance (CVs) for these comparisons were all less than 30%. These data indicate completed assay plates can be stored for up to 30 days at –20 °C before reading.

Precision

Intra-assay variability

Intra assay variability for unstimulated control and PHA stimulated samples was determined by assaying eight replicates of whole blood samples from six healthy volunteers in a single assay. The mean, standard deviation (SD)

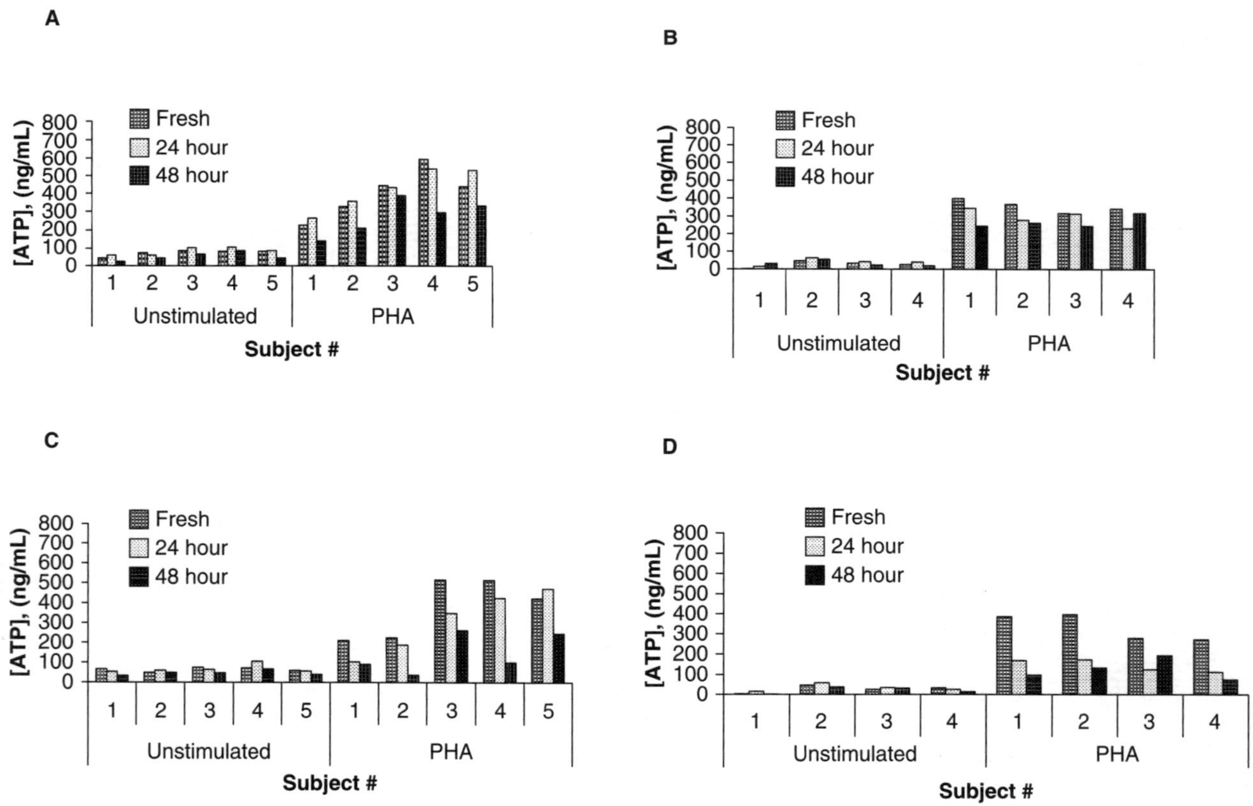

Figure 12.2 Shipping conditions. Whole blood samples from 5 subjects were shipped at either ambient temperature (*A* Cylex CD19 kit; *B*, Cylex CD4 kit) or with ice packs (*C* Cylex CD19 kit; *D*, Cylex CD4 kit) at 0, 24 and 48 hours post draw, stimulated upon arrival with PHA (2.5 μg/mL) overnight and analyzed using the appropriate Cylex kit.

and %CV were calculated for each of the six donors. The pooled %CVs for the unstimulated and PHA activated samples were below 20% in every case.

Inter-subject variability

Inter-subject variability was assessed using whole blood samples from 20–21 healthy volunteers (10 males, 10–11 females); each donor sample was assayed in quadruplicate. Increases in ATP production were measured in CD4 cells in response to PHA and LPS or in PHA, SEA and SEB in CD19+ cells and compared to production in the appropriate unstimulated cells. All 21 donors demonstrated an activation response (over unstimulated control) when stimulated with PHA (Figure 12.3), however a minimal response was measured in cells stimulated with LPS, SEA or SEB (data not shown). In

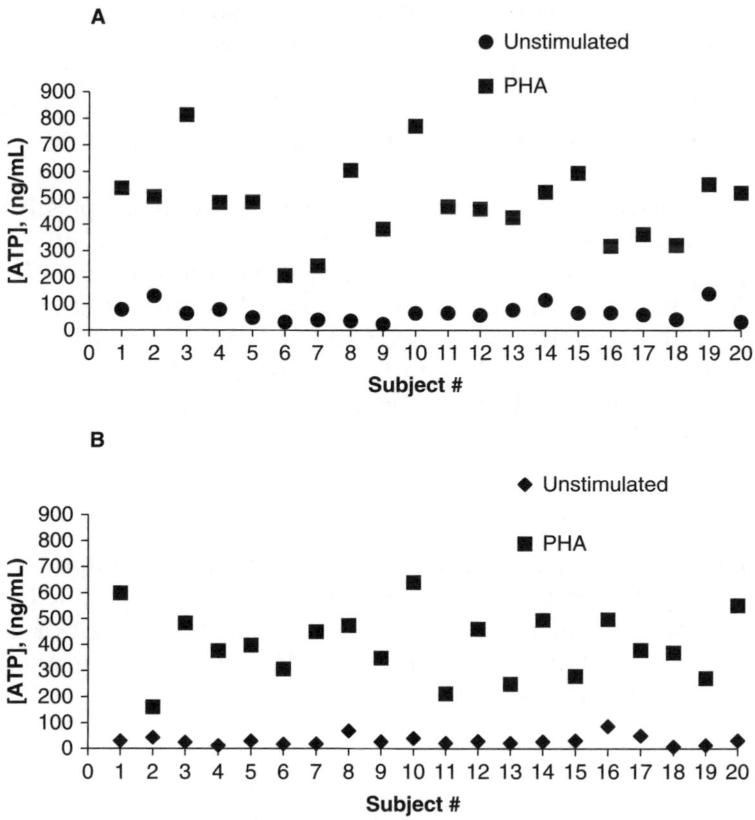

Figure 12.3 Inter-subjects variability. Whole blood samples from 21 (*A*, Cylex CD19 kit), or 20 (*B*, Cylex CD4 kit) healthy subjects were stimulated with PHA (2.5 μg/mL) overnight and assayed in quadruplicate using the appropriate Cylex kit.

the CD19 assay, the ATP concentrations of unstimulated, PHA, and LPS stimulated cells ranged from 23–137 ng/mL, 206–813 ng/mL, and 30–118 ng/mL respectively.

In the CD4+ assay, the ATP concentrations of unstimulated, PHA, SEA and SEB stimulated cells ranged from 12–86 ng/mL, 160–639 ng/mL, 41–192 ng/mL and 35–260 ng/mL respectively.

Given the low stimulation response detected with LPS, SEA and SEB, these two antigens were not part of the validation of these assays.

Inter-assay variability (inter-plate and inter-day)

Inter-plate variability was assessed using whole blood samples from four to five healthy volunteers. Donor samples were assayed in quadruplicate on three separate plates each set up on the same day independently of each other. For each donor, mean ATP values for PHA stimulated samples and unstimulated controls were calculated and compared across plates. In the case of CD19, pooled %CV for all five donors was 9.45% for unstimulated samples and 10.24% for PHA stimulated samples. In the CD4 assay, the pooled %CV for all five donors was 85.65% for unstimulated samples and 9.59% for PHA stimulated samples. The results of this evaluation are shown in Figure 12.4.

Inter-day variability was assessed using whole blood samples from four healthy volunteers, assayed in quadruplicate on at least six different days spanning a period of at least 111 days. For each donor, mean ATP values for PHA stimulated samples and unstimulated controls were calculated and compared across assay days. Pooled %CV values were calculated as 20.49% and 32.71% for CD19 and CD4 cells respectively, in the unstimulated donor samples and 25.61% and 27.46% for CD19 and CD4 PHA stimulated samples, respectively. The results are shown in Figure 12.5.

Inter-operator variability

Inter-operator variability was assessed using whole blood samples from four to five healthy donors assayed on the same day by two qualified analysts. Each operator performed the assay using a separate assay kit of the same lot and a separate aliquot of whole blood. Pooled %CV remained below 20% for all samples in both assays.

Lot variability

Inter-lot variability was assessed using whole blood samples from five to six healthy volunteers. The donor samples were evaluated on the same day in two different lots of CYLEX kits. Minimal variability was observed with pooled %CV values remaining below 25% for all samples in both assays.

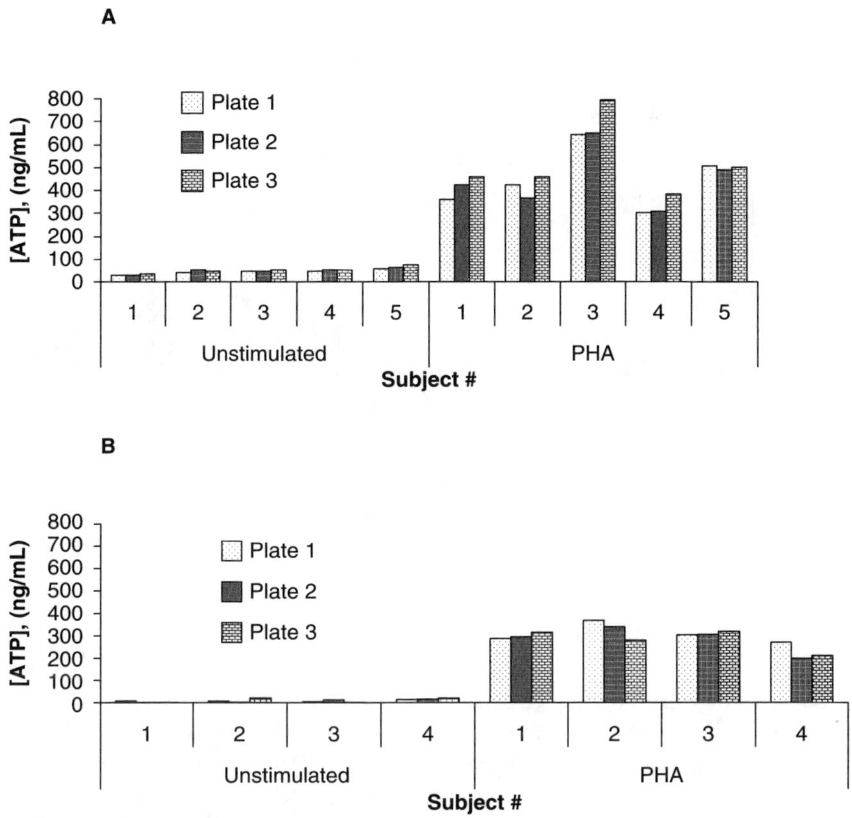

Figure 12.4 Inter-plate variability. Whole blood samples from 5 (*A*, Cylex CD19 kit), or 4 healthy subjects were stimulated with PHA (2.5 μg/mL) overnight and assayed in quadruplicate in each of three separate plates set up on the same day, using the appropriate Cylex kit .

Specificity

Drug interference

Interference or enhancement of the activation response, due to the presence of therapeutic antibodies, was evaluated by measuring the change in activation of six normal donor blood samples stimulated with PHA in the presence of no IgG, irrelevant IgG or a number of Centocor proprietary antibodies. None of the antibodies tested had an effect on PHA-induced ATP production in vitro (Figure 12.6). Since the presence of these drugs in our clinical samples would not interfere with the assay format directly, a decreased ATP response in subjects treated with CNTO products would be interpreted as a decreased immune response *in vivo*.

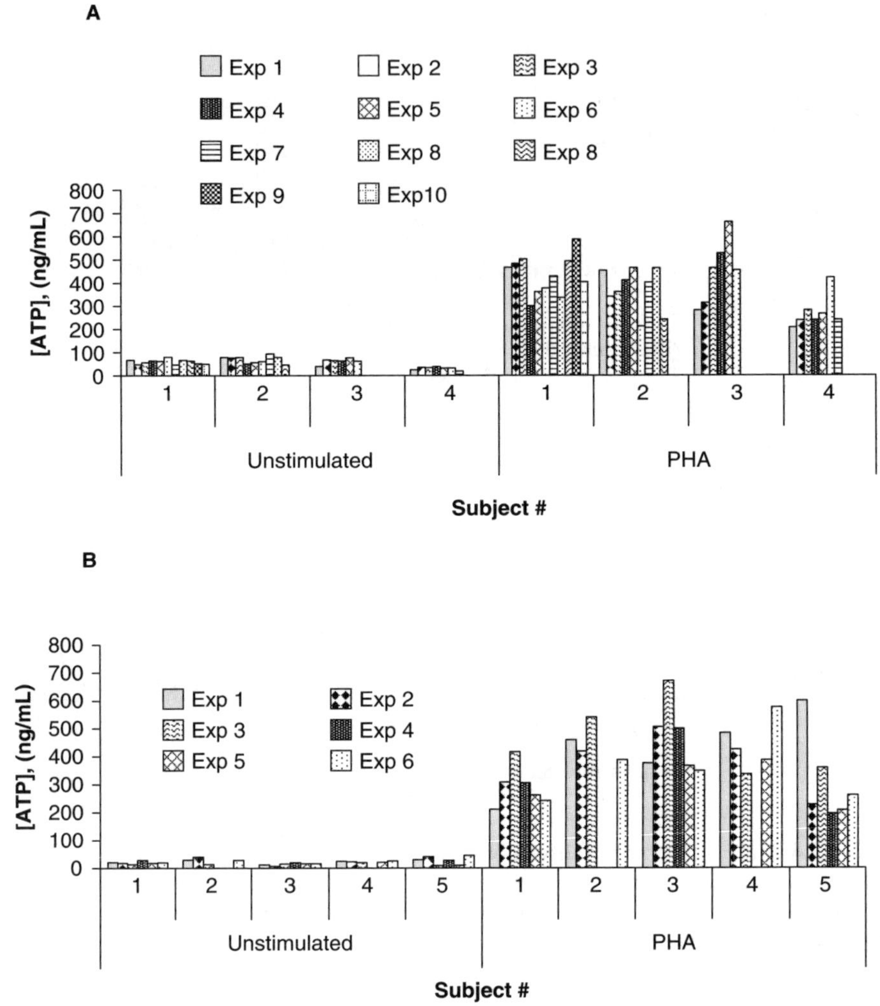

Figure 12.5 Cylex CD 19 inter-day variability. Whole blood samples from 4 (*A*, Cylex CD19 kit) or 5 (*B*, Cylex CD4 kit) healthy subjects were stimulated with PHA (2.5 μg/mL) overnight and assayed in quadruplicate using the appropriate Cylex kit in ten separate experiments, spanning 154 days (*A*, Cylex CD19 kit) or six separate experiments spanning 111 days (*B*, Cylex CD4 kit).

Assay performance and acceptance criteria

As a result of the previous validation experiments, the following acceptance criteria were established for evaluation of samples in the clinical setting.

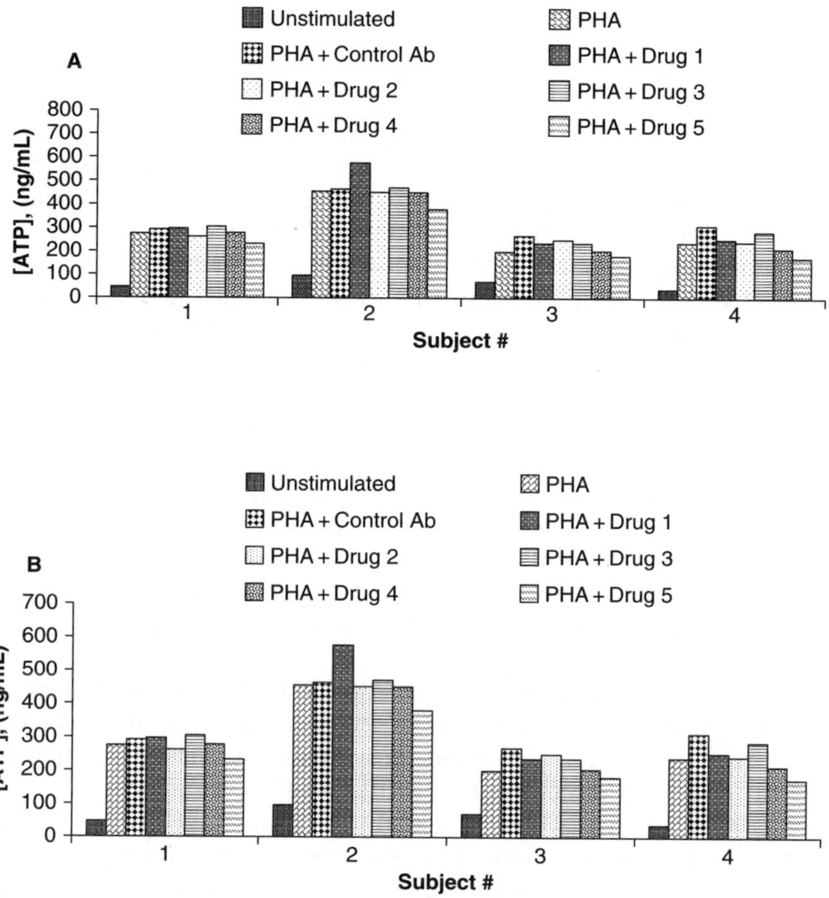

Figure 12.6 Drug interference. Whole blood samples from 4 healthy subjects were stimulated overnight with PHA (2.5 μg/mL) alone, in the presence of control IgG (10 μg/mL) or five Centocor proprietary antibody-based drugs (10 μg/mL) and assayed using the appropriate Cylex kit (*A* Cylex CD19 kit; *B*, Cylex CD4 kit).

Calibration curve

The linearity of the calibration curve must be accepted under the following criteria based on recommendations from the manufacturer. The calculated value for the 1000 ng/mL ATP Calibrator must be 900–1100 ng/mL. The calculated value for the 100 ng/mL ATP Calibrator must be 85–115 ng/mL. The correlation coefficient (r^2) of the ATP Calibration Curve must be greater than 0.97.

Non-stimulated sample

In the CD19 assay, the mean ATP concentration for non-stimulated samples must be less than 130 ng/mL. In the CD4 assay, we have adopted the criteria recommended by the manufacturer and approved by the FDA. Therefore, the mean ATP concentration for non-stimulated samples must be less than or equal to 60 ng/mL. If the specimen meets the criteria for the non-stimulated sample, proceed with interpretation of the stimulated sample. If the non-stimulated sample does not meet these criteria, the specimen cannot be evaluated but since the sample will have been used, it will not be repeated. Sample will be reported as not tested (NT)

For experimental sample

The Coefficient of Variance for experimental samples, that is those stimulated with PHA, must be less than 30%. The samples need to be evaluated in quadruplicate to account for the inherent variability of cell-based assays. The 30 %CV may be evaluated using an outlier formula. Remove the suspected outlier (the value furthest from the mean) and recalculate the mean and SD of the triplicate results.

While we can quantify the levels of ATP in these cell populations, these assays remain qualitative in nature given the high variability observed in immune response levels of normal individuals. In such qualitative assays the result does not directly quantify the level of immunosuppression or immunoactivity. While an ATP value may be given, levels will only provide guidance in stratifying patients into two categories: "normal" and "low" responders. The data then should be reported as "normal" immune response or "low" immune response. In the case of CD19, the mean ATP concentration of the same subject sample, stimulated with antigen must be double the non-stimulated sample value to be considered "normal". ATP levels less than double the non-stimulated sample value are qualified as a "low" immune response and may indicate an effect of the drug on the subject's immune system. In the case of CD4, ATP level for experimental sample must be greater than or equal to 225 ng/mL to be considered "normal". ATP levels below 225 ng/mL are qualified as a "low" immune response and may indicate an effect of the drug on the subject's immune system. Once again, we have used the work already completed by the manufacturer in the transplantation patient population as a guideline for our acceptance criteria (Kowalski et al, 2003).

In order to maintain the validated status of this assay, regular testing of additional manufacturing lots of the kits and samples should be ongoing. In addition, if the manufacturer makes any changes in the format of the assay or on any of the components of the kit, these new variables should be evaluated to maintain a regulated environment for the data. The magnitude of the change will determine the extent of the revalidation experiments required.

Furthermore, as the assay is employed to evaluate different patient populations in support of new clinical programs, blood samples from that patient population should be evaluated to reestablish cut-off values for the unstimulated samples as well as expected range of results for the experimental samples. These additional evaluations can be added as addendums to the validation here presented.

References

FDA (2001). Department of Health and Human Services. Food and Drug Administration.

Hooper E, Hawkins DM, Kowalski RJ, Post DR, Britz JA, Brooks KC and Turman MA (2005). *Clin Transplant*, **19**, 834–849.

Keilholz U, Weber J, Finke JH, Gabrilovich DI, Kast WM, Disis ML, Kirkwood JM, Scheibenbogen C, Schlom J, Maino VC, Lyerly HK, Lee PP, Storkus W, Marincola F, Worobec A and Atkins MB (2002). *J Immunother*, **25**, 97–138.

Knight RJ, Kerman RH, McKissick E, Lawless A, Podder H, Katz S, Van Buren CT and Kahan BD (2005). *Transplant Proc*, **37**, 3538–3541.

Kowalski R, Post D, Schneider MC, Britz J, Thomas J, Deierhoi M, Lobashevsky A, Redfield R, Schweitzer E, Heredia A, Reardon E, Davis C, Bentlejewski C, Fung J, Shapiro R and Zeevi A (2003). *Clin Transplant*, **17**, 77–88.

Paul W (2003). *Fundamental Immunology*, Lippincott - Raven, Philadelphia.

Sottong PR, Rosebrock JA, Britz JA and Kramer TR (2000). *Clin Diagn Lab Immunol*, **7**, 307–311.

Zeevi A, Britz JA, Bentlejewski CA, Guaspari D, Tong W, Bond G, Murase N, Harris C, Zak M, Martin D, Post DR, Kowalski RJ and Elmagd KA (2005). *Transpl Immunol*, **15**, 17–24.

13 Development of validated neutralization bioassays

Manoj Rajadhyaksha[1], Manjula Reddy, Jaime Bald, Amy Fraunfelter, Persymphonie Miller, Marian Kelley and Uma Prabhakar

[1] Centocor Research and Development, Inc., 145 King of Prussia Road, Radnor, PA 19087, USA

13.1 Introduction

The development of anti-drug (protein therapeutic molecule) antibodies (ADA) has emerged as an important safety concern for drug manufacturers, clinicians and patients (Rosenberg and Worobec, 2004; Science Module; Schellekens, 2002). The safety concerns range from adverse events like infusion reactions and hypersensitivity to effects on drug exposure, resulting in faster clearance rate and altered biodistribution of the drug.

* To whom correspondence should be addressed.

Validation of Cell-Based Assays in the GLP Setting: A Practical Guide Editors, Uma Prabhakar and Marian Kelley

At times, cross-reaction of ADA with the analogous endogenous host derived protein can lead to a fulminant autoimmune response attributed to ADA being neutralizing antibodies (NAbs). Observations like pure red cell aplasia associated with anti-erythropoetin antibodies in patients treated with recombinant erythropoietin underscores the importance of monitoring for NAbs in subjects administered with any biologic therapeutic (Casadevall et al, 2002). Consequently, The United States Food and Drug Administration (FDA) now requests for detailed analysis plans for characterizing anti-drug antibody responses especially in late phase clinical trials intended for registration purposes (Wadhwa et al, 2003).

Typically ADAs are detected and analyzed by enzyme-linked immunoassays (EIAs) by a variety of technology platforms. The general principle behind these EIAs involves the binding of antibodies to the drug followed by ultra-sensitive signal amplification using various approaches. However, these technologies only provide an idea of the presence of NAbs, but do not characterize the biology of ADA. Functional cell based bioassays is perhaps the only approach that can be used to characterize NAbs and provide some measure of their activity in vivo. Bioassays generally involve the use of a responsive cell-line responding to a target molecule. The interaction of the target with the cell-line leads to a biological event that can be monitored and measured. The outcome of such an interaction could be cell growth, cell death, or secretion of a cytokine or protein factor. If NAbs are present, they interfere with the binding of the drug and the target molecule and there by interfere with the biological outcome of the target and cell interaction.

Bioassays are difficult to standardize, expensive, labor intensive, and sometimes lack sensitivity, reproducibility and even precision. Therefore a bioassay designed to detect neutralizing anti-therapeutic antibody response should strike the right balance between being as close as possible to reflect the intended use of the biologic but at the same time be sensitive, validatable and cost-effective (Mire-Sluis, 2001; Wei et al, 2004).

Here we report the validation of two different types of neutralization bioassays, one in which the drug (DRUG-X a biologic monoclonal antibody) is directed against Tumor necrosis factor (TNF) with cell killing as the eventual read out and the second, in which the drug (DRUG-Y a biologic monoclonal antibody) inhibits cytokine IL-12-induced Interferon-gamma (IFN-γ) secretion by NK cells with IFN-γ being the final read out. While there are no specific guidelines for the validation of such NAb bioassays, recently Gupta et al (2007) have provided recommendations on the development, optimization, and qualification of non-functional cell-based assays for assessing the neutralizing capacity of anti-drug product antibodies by using a fixed concentration of drug product in the NAb assay. Another publication (Mire-Sluis et al, 2004) provides scientific recommendations based on the experience of the authors for the development of anti-product antibody immunoassays intended for preclinical or clinical studies. The focus of this publication is on assay design considerations, and scientific background to the various assay performance parameters necessary for developing a valid assay. Here, we have followed the general guidelines (best practice approach) presented in the ICH guidelines (ICH, 1997) for developing and validating functional cell-based assays for NAbs.

13.2 NAb Assay Development

Principle of the assay

WEHI cells apoptose in a dose dependent manner in the presence of TNFα. TNFα is neutralized by the drug (DRUG X), which in turn can be neutralized by anti-DRUG X. Hence at a fixed concentration of TNFα and DRUG X in the assay well, the presence of anti-DRUG X leads to neutralizing of the DRUG X drug and thereby the killing of the WEHI cells by non-neutralized TNFα in an anti-DRUG X dose dependent manner. The viability and functionality of the cells is quantitated by the conversion of a metabolic reagent (substrate–CellTiter-Glo™) that emits light (chemiluminescence), which is measured by a luminometer. Thus, higher levels of anti-DRUG X in the sample lead to lower relative luminescence unit (RLU) values. From the RLUs as the source data, percent inhibition of the bioactivity of the DRUG X drug is computed and presented as the final modified data.

Controls for the DRUG X assay

There are three controls designed to monitor the validity of the DRUG X bioassay designated as *Assay Controls*, *Consistency Controls* and *Sample Specific Controls*.

Assay Controls are run once per assay and consist of the following six controls: (1) Media: monitors the background signal contributed by the metabolic reagent (CellTiter-Glo™); (2) Media+cells: monitors the overall metabolic health of the cells; (3) TNFα+cells and (4) DRUG X+cells: monitor any toxicity or proliferative potential (on cells) of these key protein reagents; (5) anti-DRUG X1+cells and (6) anti-DRUG X2+cells: (DRUG X1 and DRUG X2 are anti-DRUG X low and high affinity antibodies respectively) monitor any toxicity or proliferative potential (on cells) of these key protein reagents.

Consistency Controls composed of both negative and positive controls, monitor the consistent performance of the bioassay and are analyzed on every assay plate. The *Consistency Controls* were prepared from two selected positive control antibodies (anti-DRUG X1 and anti-DRUG X2). The positive *Consistency Controls* monitors the drug (DRUG X) neutralization capacity of two selected positive control antibodies and the results are represented as percent inhibition (by the *Consistency Controls)* of protection offered by the drug (DRUG X) to the cells against TNFα. The final assay well components of the positive *Consistency Controls* were composed as follows: Anti-DRUG X1 (a low affinity mouse monoclonal antibody anti-DRUG X)+DRUG X+TNFα+ cells (pCCX1) & Anti-DRUG X2 (a high affinity mouse monoclonal antibody anti-DRUG X)+DRUG X+TNFα+cells (pCCX2). The "p" denotes positive control, and CC is consistency controls. Controls were prepared in 5% NHS (pCCX1-NHS & pCCX2-NHS). Besides these two positive consistency controls, two other negative controls were monitored. They were 5% pooled naïve normal human serum (NHS) (nCC-NHS) and a negative control antibody DRUG A (non-neutralizing antibody) in 5% NHS (nCCA). The "n" denotes the negative controls and CC is consistency controls. The nCC-NHS control monitors the baseline sensitivity of cells in each assay towards any normal serum component, where as the nCCA control demonstrates the inability of a related (non-neutralizing) antibody to inhibit the DRUG X bioactivity.

Sample Specific Controls are used to monitor the influence of each individual serum on the cells alone, as well as for any interference of TNFα killing of the cells. Two individual *Sample Specific Controls* consisting of 5% test serum or plasma sample+cells, and 5% test serum or plasma sample+cells+TNFα is evaluated (for every serum or plasma sample tested during the validation) to monitor the influence of each sample on the cells as well as on the baseline TNFα cytotoxicity.

To be valid, all the *Controls* had to qualify as per their acceptance criteria.

Assay procedure

Test samples and controls were diluted in media at a 1/20 dilution. DRUG X1 and DRUG X2 controls were prepared by spiking 20 fold excess concentration of the respective antibody. The diluted controls and samples were plated at 25 μL/well in a 96-well microtiter plate. DRUG X (at a predetermined concentration) was added at equal volumes to the control and sample wells. The plates were then incubated at 37 °C with 5% CO_2 for 2 hours. Next, diluted TNFα at pre-standardized concentrations was added and the plates incubated at 37 °C with 5% CO_2 for 2 hours. WEHI-C527 cell at 2×10^6 cells/mL with 4 μg/ml Actinomycin D–Mannitol (Sigma catalog number A-5156) were added to wells containing DRUG X anti-DRUG X and TNFα, in a volume of 25 μL. The assay plate was then incubated for 24 hours (± 4 hours) at 37 °C with 5% CO_2. Following 18–20 hours, the assay plate was pulsed with 100 μL/well of CellTiter Glo reagent (Promega catalog number G7570). The luminescence was read on a SpectraMax M2 reader.

Results were calculated as percent inhibition of the protection offered by the drug against TNFα killing. The percent inhibition results were calculated from relative fluorescence units (RLU) as described:

$$\% \text{ Inhibition formula} = (1 - (\text{sample RLU/TNF} + \text{DRUG X} + \text{cells RLU}) \times 100)$$

TNF+DRUG X+cells is a control that allows the measurement of total protection offered by DRUG X (drug) to the WEHI cells against TNFα.

Determination of an assay cut-off

Two individual assay cutoffs, one for human (inclusive of serum and plasma) and the other for primates (inclusive of serum and plasma) were established. Two operators tested serum and plasma samples collected from a naïve population of human or primate subjects. Naïve subjects, as defined here, included those subjects who had never been exposed to DRUG X. The mean percent inhibition of DRUG X bioactivity and SD for serum and plasma samples was computed. Based on mixed-effect analysis and taking into account random effects like samples, dates, and plates a cumulative analysis was performed for all primate serum and plasma samples tested to generate a primate assay cut-off of +10.2% inhibition (SE of 16.7%) of DRUG148 bioactivity.

Similarly an assay cut-off for human samples using 24 healthy normal human subjects and naïve disease populations from autoimmune diseased and cancer patients was established. EDTA and heparinized plasma samples were tested from a subset of ten of these 24 normal human sera. The mean percent inhibition result for serum samples in this group was –18.01%, whereas the analysis of the ten naïve human EDTA plasma samples resulted in a mean percent inhibition result of 17.82%. The heparinized plasma sample analysis resulted in a mean percent inhibition result of –11.24%. Based on these observations it was evident that the EDTA plasma and heparin plasma samples may behave differently. All EDTA plasma samples demonstrated an artifact of false positive percent inhibition results even though they do not contain any inhibiting anti-DRUG148 antibodies. Therefore the EDTA plasma samples were eliminated from the human cut-off analysis.

Serum and heparinized plasma samples were grouped to calculate the assay cut-off. Analysis for similarities in serum and plasma and diseased populations was performed as follows:

$$\begin{aligned} &\ln(1-\%\text{inhibition}/100) = \ln[(\text{Mean RLU of TNF+DRUG X+cells})/ \\ &\quad (\text{Mean RLU of 1:20 sample+DRUG X+TNF+cells})] = \\ &\ln(\text{Mean RLU of TNF} + \text{DRUG X} + \text{cells}) - \\ &\ln(\text{Mean RLU of 1:20 sample} + \text{DRUG X} + \text{TNF} + \text{cells}) \end{aligned}$$

where ln is natural log.

Based on a mixed effects analysis, there was no significant difference among the adjusted mean inhibitions of the three types of samples (normal serum, diseased serum, and normal plasma). With a p value of 0.87 the samples were grouped using the type of sample and disease as fixed effects and adjusted (accounted) for the following random effects: Samples, dates, plates, and date by disease and type interaction.

A cumulative analysis of human samples resulted in a cutoff of +6.45% (SE 2.75%) inhibition ($0.1455 - z \times 0.1290 = -0.0667$ on a log scale) using $z = 1.645$ to produce a 5% false positive rate taking into account the current industry opinion where an immune response assays should have a false positive rate of approximately 5% (Mire-Sluis et al, 2004; Geng et al, 2005). In this analysis there were no fixed effects. The only random effects were samples and plates. Other potential random effects like dates, operators, and various interactions, were negligible and non-significant. Thus a human sample (heparin plasma or serum) percent inhibition result greater than or equal to +6.45% is considered positive in this anti-DRUG148 IR bioassay.

Assay sensitivity

The anti-DRUG X NAb assay sensitivity was established by testing serial two-fold dilutions covering a range from 1248 to 0.625 ng/mL of a known high affinity anti-DRUG X neutralization positive antibody (DRUG X2), in multiple assays (Figure 13.1). The dilution curve results were compared

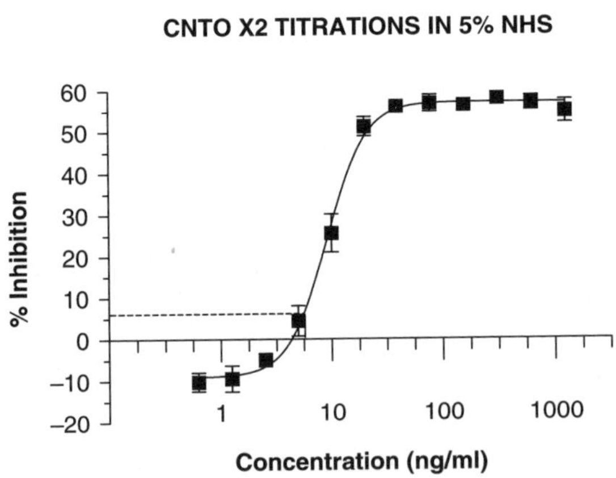

Figure 13.1 Assay sensitivity was established as the lowest detectable concentration of antibody relative to the +6.45% inhibition cut-off (dotted line). The neutralization curve of DRUG X2 depicted in this figure is based on the mean of five independent analyses

to the assay cut-off (+6.45% inhibition) and the lowest detectable concentration of the DRUG X2 antibody was determined to be 7.5 ng/mL. Thus the assay sensitivity based on the evaluation of this monoclonal DRUG X2 antibody was 150 ng/mL based on sample dilution (1/20).

Validation of controls

In order to monitor the validity of the DRUG X Nab bioassay acceptance criteria were established for *Assay Controls*, *Consistency Controls* and *Sample Specific Controls*. From the *Assay Controls* an important parameter called *Protection Factor* (TNF + DRUGX + cells/TNF + cells) was calculated. The protection factor which provides an index of sensitivity and provides an assessment of linearity was established to be greater than 1.44.

Consistency controls

Two neutralization positive antibodies, a high affinity antibody (DRUG X2-in 5% normal human serum) and a weak affinity antibody (DRUG X1-in 5% normal human serum), were selected to demonstrate that this assay is capable of detecting high affinity as well as moderate to weak affinity antibodies. Acceptance ranges (M+3SD) of the percent inhibition results for both the negative and positive consistency controls were established.

Sample Specific Controls

Samples from normal individual sera ($n = 24$) and a variety of diseased ($n = 10$) population were used to establish the acceptance criteria for sample specific controls.

Assay specificity

To test for assay specificity, unrelated antibodies like an anti-anti-IL-6 antibody, anti-anti-IL-12 antibody, anti-$\alpha V\beta 3$ integrin antibody and, anti-anti-erythropoietin mimetobody were tested in the anti-DRUG X NAb Bioassay. The concentrations were selected to cover the concentrations at which the consistency controls (High and Low affinity) were formulated. Results indicated that the assay was specific for the detection of neutralizing anti-DRUG X antibodies, and interference from other unrelated antibodies was minimal.

Drug interference limit

To demonstrate the drug (DRUG X) interference limits of the assay, increasing DRUG X concentrations were spiked into 5% pre-diluted NHS containing 40 ng/mL of pre-spiked DRUG X2 (positive control antibody). Figure 13.2 demonstrates that with increasing concentration of DRUG X the ability of the spiked DRUG X2 antibody to neutralize DRUG X in the assay mixture is inversely proportional to the DRUG X concentration. The maximum interfering concentration of exogenous DRUG X was

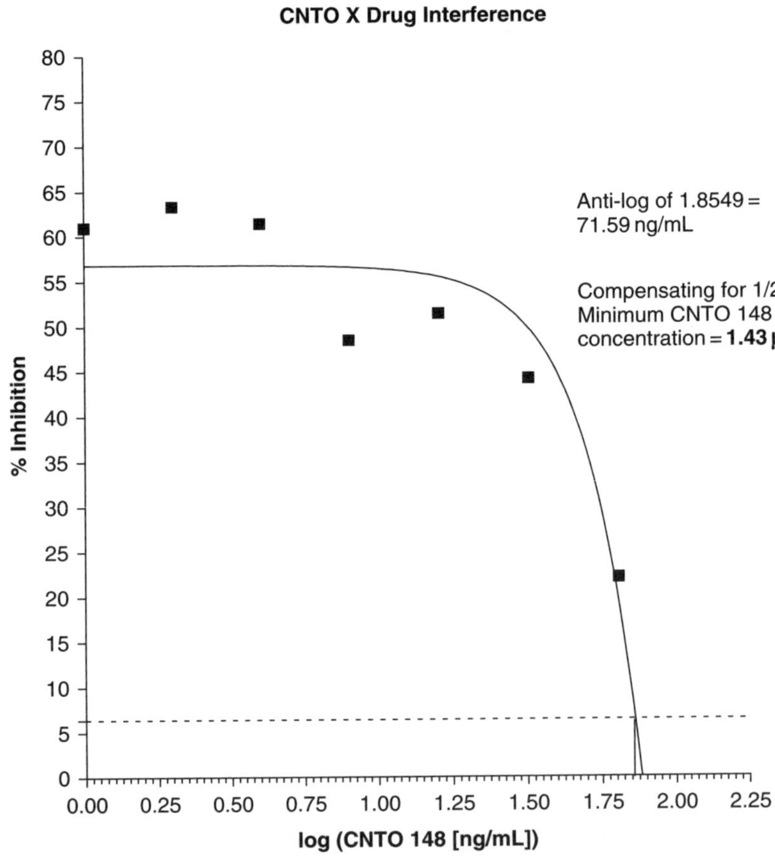

Figure 13.2 DRUG X at increasing concentrations was spiked into a 1/20 pre-diluted NHS containing 40ng/mL of pre-spiked DRUG X2 (high affinity anti-DRUG X antibody). The mock samples thus prepared mimicked the conditions as observed in a serum sample with interfering DRUG X concentrations in the presence of anti-DRUG X antibodies. These mock samples were then treated as per the assay method and tested in the Bioassay. 1.43μg/mL of DRUG X was observed as the least concentration required for total (result below the assay cut-off of +6.45%) interference in the assay

interpolated to be 71.59 ng/mL, after correcting for the 1/20 assay dilution it translates to 1.43 μg/mL of DRUG X, which can totally abolish (below the assay cut-off of +6.45%) the neutralization capacity of a high affinity anti-DRUG X antibody at 40 ng/mL in the sera.

Robustness

Assay robustness and consistency for the bioassay were analyzed as follows: (a) pre-incubation of DRUG X and anti-DRUG X, (b) secondary incubation of TNF with the pre-incubated DRUG X+anti-DRUG X mixture and, (c) incubation of the detection system Cell Titer Glo (10 minutes).

Two steps of the anti-DRUG X NAb Bioassay that were tested were as follows: Step A, DRUG X +anti-DRUG X; Step B, TNF+(DRUG X+ anti-DRUG X) mixture. Under one experimental condition the Step A incubation was increased by 1 hour maintaining the Step B similar to the standard conditions. In the second experimental condition the Step B incubation was increased by 1 hour while maintaining the Step A similar to the standard conditions. Results were compared with the standard conditions (2 hours per step) of incubation. All percent inhibition results obtained for the Consistency Controls (under both the changed incubation conditions) and Assay Controls were consistent and comparable to the results obtained with standard conditions. Percent recovery was calculated from the percent inhibition results of the positive consistency controls. The percent recovery of the inhibition for positive controls under the changed conditions ranged from 86.76% to 104.71% (increasing by 1 hour) or ranged from 89.72% to 107.47% (decreasing by 1 hour), indicating that the assay was rugged under these conditions.

Assay Ruggedness was also evaluated following the 10 minute incubation step of Cell TiterGlo. After addition of the Cell Titer Glo solution to the assay wells the cells are allowed to lyse for 10 minutes and the released ATP molecules are quantitated luminometrically. The incubation time was gradually increased in increments of 10 minutes up to 30 minutes. Results were compared with those of the standard condition. The percent recovery of the inhibition for positive controls under the changed conditions ranged from 99.81% to 106.69%. Overall, the anti-DRUG X Nab bioassay was shown to be rugged.

Reproducibility

To assess assay reproducibility, two operators performed multiple analyses of the positive consistency controls. The assay reproducibility was evaluated by testing for inter-, intra-assay, and operator-to-operator variation. Reproducibility of results between different batches of WEHI cells and

Table 13.1 Intra-assay variation obtained by a single operator on WEHI-C527 cells by testing the two positive consistency controls in the anti-DRUG X Nab assay. Results expressed as percent inhibition of the DRUG X bioactivity

Consistency controls	PCCX1- NHS	PCCX2- NHS
	52.34	72.46
	53.52	73.79
	51.28	74.85
	54.6	73.89
	51.28	71.26
	57.59	73.66
	48.9	72.54
	52.23	69.52
Mean	52.78	72.49
SD	2.43	1.77
%CV	4.61	2.44

influence of the passage numbers on the obtained results was also analyzed as part of the reproducibility evaluation.

Table 13.1 shows intra-assay variation obtained by a single operator in eight repetitions of the positive consistency controls in a single assay. Mean, SD and the intra-assay variation (%CV) were calculated. The intra-assay variation for the two positive consistency controls ranged from 2.44% to 4.61%.

As shown in Table 13.2, two operators using three different passage numbers of the WEHI-C527 cells tested the two positive consistency controls in the anti-DRUG X Nab assay. A total of six assays by two operators (three by each operator) were analyzed and the inter-assay variation was calculated. The inter-assay variation for the two positive consistency controls between these two operators ranged from 8.68% to 14.60%. The inter-operator coefficient of variability (%CV) for all four positive consistency controls ranged from 2.42% to 3.63%.

A single operator evaluated assay variability across different cell passage numbers (passage numbers 3, 5, 7, 13, and 16). The two positive consistency controls two negative consistency controls and assay controls were tested on each of the passages of these cells. For the two positive consistency controls the %CV ranged from 3.0% to 5.60%. These data indicate that assay reproducibility is consistent and within acceptable limits up to passage 17.

Lot-to-lot comparison of reagents

Lot-to-lot variability was assessed with reference to critical reagents such as, TNF, DRUG X, Actinomycin-D and Cell titer Glo. Two different lots of TNF were tested on the WEHI-C527 cells. The EC-50 of the two TNF lots

Table 13.2 Inter-assay variation and operator-to-operator variation obtained by two operators on C527 cells by testing the two positive consistency controls in the anti-DRUG X b assay. Results expressed as percent inhibition of the DRUG X bioactivity

Inter-assay variation									
Operator	Operator 1			Operator 2					
Passage number	8	12	17	8	12	17			
Consistency controls							Mean	SD	%CV
PCCX1-NHS	55.11	43.00	36.99	43.4	39.4	47.78	44.28	6.46	14.60
PCCX2-NHS	72.63	68.14	61.67	70.3	57.3	64.69	65.79	5.71	8.68

Inter-operator variation

Consistency controls	Mean	SD	CV %
PCCX1- NHS	44.28	1.07	2.42
PCCX1- NHS	65.79	2.39	3.63

was 18.43 and 19.27 pg/mL respectively. Similarly, reproducibility within two individual lots of DRUG X, Actinomycin-D mannitol (Act-D) and three lots of Cell Titer Glo was fairly comparable.

Stability of consistency controls

Stability of critical reagents such as, Actinomycin-D, TNF and the consistency controls at 4 °C and –70 °C stability was evaluated. The reconstituted TNF solution was found to be stable for at least 14 days at 4 °C. All the Consistency controls and Assay controls performed within expectations and the results obtained on all controls were very comparable to those using TNF prepared fresh. Actinomycin-D was stable for 14 days.

Stability of the positive consistency controls at 4 °C (as the stress condition) was evaluated for 28 days (Table 13.3). At the initiation of these experiments, four individual sets of consistency controls were stored at –70 °C. Tubes of each set were composed of prepared and frozen positive and negative consistency controls. The first set of controls was thawed and stored at 4 °C for 28 days. The second and third sets were thawed on a weekly basis and stored for day 21 and day 14 time points. Finally the day 0 time point was obtained by thawing the last set of controls on the day of the testing. On the day of testing, the test consistency controls were

Table 13.3 Stability of positive consistency controls at 4 °C for 28 days. Results are reported as percent inhibition (Consistency Controls) of the DRUG X bioactivity. Percent recovery of the inhibition observed for freshly prepared controls is indicated in parenthesis

Days at 4 °C	Freshly prepared %Inhib	Day 0 %Inhib	Day 14 %Inhib	Day 21 %Inhib	Day 28 %Inhib	Mean	%CV
PCCX1-NHS	37.97	43.12 (113.55%)	37.93 (99.90%)	35.12 (92.49%)	35.06 (92.34%)	37.84	8.66
PCCX2-NHS	58.18	58.92 (101.27%)	61.05 (104.93%)	59.17 (101.69%)	60.22 (103.50%)	59.51	1.90

Table 13.4 Stability of positive consistency controls at –70 °C for 28 days. Results are reported as percent inhibition (Consistency Controls) of the DRUG X bioactivity. Percent recovery of the inhibition observed for freshly prepared controls is indicated in parenthesis

Days at –70 °C	Freshly prepared %Inhib	Day 14 %Inhib	Day 21 %Inhib	Day 28 %Inhib	Mean	%CV
PCCX1-NHS	37.97	33.28 (87.65%)	39.07 (102.89%)	43.97 (115.79%)	38.57	11.37
PCCX2-NHS	58.18	34.97 (60.11%)	47.14 (81.02%)	60.99 (104.83%)	50.32	23.60

compared against freshly prepared positive consistency controls. Results are reported as percent inhibition of the DRUG X bioactivity (Table 13.3). As seen in Table 13.3 the variation of controls ranged from 1.90% to 8.66% and the corresponding percent recovery of the inhibition under the stability study conditions ranged from 92.34% to 113.55%. Similarly, controls stored at –70 °C (Table 13.4) were shown to be stable when stored at –70 °C for up to 28 days.

The effect of multiple freeze–thaw cycles on the positive consistency controls was evaluated. As shown in Table 13.5 the %CV of the percent inhibition results ranged from 6.15% to 13.52%, which demonstrates that up to, three consecutive freeze thaw cycles are well tolerated. Overall, the stability studies indicate that the positive and negative consistency controls are stable under the various conditions tested.

Table 13.5 Stability of positive consistency controls after up to three freeze–thaw cycles. Three sets of consistency controls were frozen at –70 °C and were thawed and refrozen for one, two or three freeze–thaw cycles. Stability of the frozen and thawed controls was compared to freshly prepared controls. Results are reported as percent inhibition (Consistency controls) of the DRUG X bioactivity. Percent recovery of the inhibition observed in comparison to freshly prepared controls is indicated in parenthesis

Consistency Controls	Freshly prepared % Inhib	One freeze–thaw cycle % Inhib	Two freeze–thaw cycles % Inhib	Three freeze–thaw cycles % Inhib	Mean % Inhib	CV
PCCX1-NHS	42.81	41.67 (97.34%)	43.13 (100.75%)	37.61 (87.85%)	41.30	6.15
PCCX2-NHS	67.13	49.01 (73.01%)	63.67 (94.85%)	65.19 (97.11%)	61.25	13.52

Determination of NAb reactivity in a clinical sample

In order to determine the presence of Nabs in clinical samples a percent change value from baseline cut-off was established. In order to account for serum matrix effects the mean percent inhibition for each post treatment sample was normalized with respect to its baseline (pretreatment or week 0) sample and a "percent change from baseline" was calculated as follows:

$$100 \times (\%\text{inhibition post} - \%\text{inhibition baseline})/(100 - \%\text{inhibition baseline})$$

The percent change from baseline cut-off is based on a two standard deviation limit and an intra-plate variance component of 4.71% estimated from the assay validation data. The critical value for the "percent change from baseline" results was calculated to be 13. Therefore, if a post treatment sample "percent change from baseline" value is greater than or equal to 13 then the sample is determined to be positive for the presence of NAb. When the value is less than 13, NAbs are not detectable and the sample is reported as negative.

13.3 Anti-DRUG Y neutralization bioassay development and validation

We next describe a NAb assay where the readout is the cytokine produced (IFNγ) by IL-12 induced NK (NK-92MI) cells. Addition of DRUG Y to the assay inhibits IL-12 induced IFN-γ production and presence of NAbs in the patient serum reverses the IFN-γ inhibitory effect of DRUG Y. The IFN-γ induction part of the bioassay (Part I) is followed by measuring the resulting IFN-γ levels in the cell culture supernatants by enzyme linked immunosorbent assay (ELISA) (Part II). NK-92MI cells are cultured with rhIL-12 and DRUG Y in the presence of patient sera samples that test positive for antibody response against DRUG Y as determined by a screening immunoassay. Clinical samples are determined to be positive for presence of NAbs to DRUG Y if there is ≥36% (bioassay cut-off value) recovery of the IFN-γ levels in the presence of the clinical sera sample.

13.4 Anti-DRUG Y NAb assay procedure

NK-92MI cells are seeded in culture flasks at 2×10^5 NK-92MI cells/mL and cultured for 2 days at 37±2 °C with 5±1% CO_2 and humidity of 95%. On day 2, induction plates are set up with the assay controls and sample controls described below, followed by the addition of cells (2.5×10^4/well) and rhIL-12. The total assay volume in each well is 200 μL. The final concentration of reagents is 0.05 ng/mL of rhIL-12, 100 ng/mL of DRUG Y, 100 ng/mL of NAb control or 1:1000 dilution of NAb positive sera. After 20±2 hours of culture, supernatants are collected and tested for IFN-γ levels at 1:5 dilution.

Assay controls (AC)

AC1 = Cells alone (assay background in 1:20 dilution of NHS)
AC2 = Cells+rhIL-12 (maximum IFN-γ response control of the assay with NHS)
AC3 = Cells+rhIL-12*+DRUG Y* (maximum inhibitable response in the assay)
AC4 = Cells+rhIL-12+DRUG Y*+polyclonal anti-DRUG Y serum* or a monoclonal neutralizing antibody* (maximum neutralization response control)

* refers to reagents that are pre-blocked together for 1 hour at 37±2 °C, 5±1% CO_2. All assay controls (AC1 to AC4) included 5% NHS to mimic serum amount in clinical sample test wells.

Sample controls (SC)

SC1 = Cells alone+clinical sample** (background response of the test sample)
SC2 = Cells+rhIL-12+clinical sample** (maximum IFN-γ response of the test sample that serves as the specificity control for the sample, to rule out any non-specific effects of serum factors). ** 1:20 dilution of clinical sample.

Test sample (TS)

TS = Cells+rhIL-12+DRUG Y*+clinical sample** (test condition for NAb activity of the sample)

13.5 Development and validation of IFN-γ ELISA

Different commercial ELISA kits were evaluated for IFN-γ, for the range of standard curve, reproducibility of standard curves between lots, and background response in the presence of serum. Validation of the ELISA included determination of limits of quantification of the assay, dilutional linearity, matrix effects, and stability of IFN-γ responses in stored culture supernatants (DeSilva et al, 2003; Findlay et al, 2000).

Standard curves and Lower and Upper Limits of Quantification (LLOQ and ULOQ)

The R&D Systems Quantikine Human IFN-γ Immunoassay kit with a standard curve range from 15.6 pg/mL to 1000.00 pg/mL was selected for this purpose. The standard curves generated over nine days were reproducible with a %CV of 7.36 (Figure 13.3).

The LLOQ and ULOQ were established by spiking a range of concentrations of rhIFN-γ (0.98–1000.00 pg/mL) into RPMI-1640 media and determining the recovery of the IFN-γ from the spiked samples. Mean percent recoveries ranged from 90.80 to 106.38 for concentrations of 15.6–1000.00 pg/mL and from 61.44 to 459.39 for lower concentrations of 0.98–7.80 pg/mL (Table 13.6). Concentrations between 0.98 and 15.6 pg/mL may not be

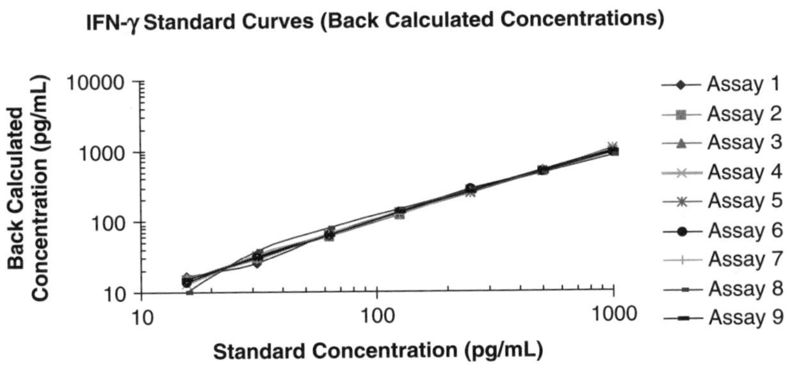

Figure 13.3 Standard curves from IFN-γ ELISA assays. Representative standard curves from nine separate IFN-γ ELISA assays are shown

Table 13.6 LLOQ determinations and spike/recovery

Spiked concentration (pg/mL)	Acceptable range	Assay 1			Assay 2			Assay 3			Mean percent recovery	Std Dev percent revovery
		Observed[a] concentration (pg/mL)	Expected[b] concentration (pg/mL)	Percent recovery[c]	Observed[a] concentration (pg/mL)	Expected[b] concentration (pg/mL)	Percent recovery[c]	Observed[a] concentration (pg/mL)	Expected[b] concentration (pg/mL)	Percent recovery[c]		
1000.00	750.00–1250.00	1070.26	1072.61	99.78	891.05	1000.00	89.10	835.22	1000.00	83.52	90.80	8.26
500.00	375.00–625.00	550.22	552.58	99.57	482.38	500.00	96.48	467.73	500.00	93.55	96.53	3.01
250.00	125.00–375.00	275.20	277.55	99.15	241.90	250.00	96.76	268.33	250.00	107.33	101.08	5.54
125.00	93.75–156.25	134.38	136.74	98.28	125.04	125.00	100.03	146.10	125.00	116.88	105.06	10.27
62.50	46.88–78.13	65.40	67.75	96.53	64.24	62.50	102.78	74.89	62.50	119.83	106.38	12.06
31.20	23.40–39.00	32.27	34.63	93.20	30.23	31.20	96.90	32.45	31.20	103.99	98.03	5.48
15.60	11.7–19.5	17.39	19.75	88.08	14.27	15.60	91.49	15.52	15.60	99.49	93.02	5.85
7.80	5.85–9.75	N/A	N/A	N/A	8.44	7.80	108.21	14.66	7.80	187.95	148.08	56.39
3.90	2.93–4.88	N/A	N/A	N/A	9.17	3.90	235.13	2.81	3.90	72.03	153.58	115.33
1.95	1.46–2.44	N/A	N/A	N/A	2.40	1.95	122.87	0.00	1.95	0.00	61.44	86.88
0.98	0.74–1.23	N/A	N/A	N/A	0.00	0.98	0.00	9.00	0.98	918.78	459.39	649.67
0.00		2.35			0.00			0.00				

[a] Observed concentration is back-calculated concentration.
[b] Expected concentration is spiked plus 0.00 pg/mL background.
[c] Percent recovery is observed/expected 100.

quantified, as the observed mean recoveries were more than 25% of the expected recovery. As results are within ±25% for 15.6 pg/mL and higher, the LLOQ was set at 15.60 pg/mL, and the ULOQ was set at 1000.00 pg/mL to concord with the concentration of standard curve calibrators. Samples with concentrations greater than 1000.00 pg/mL will have to be diluted and re-assayed.

Dilutional linearity

Dilutional linearity was determined by spiking media samples with 1000, 3000, 5000, and 10 000 pg/mL rhIFNγ and then assayed neat (undiluted) or at 1:5, 1:10, 1:20, and 1:40 dilutions. The results (not shown) indicated that concentrations below 10 000 pg/mL can be diluted 1:5 and concentrations equal to and above 10 000 pg/mL can be tested at a 1:10 dilution with a high percent recovery.

Interference of other cytokines or therapeutic antibodies in IFN-γ detection

To indicate whether or not IFN-γ in cell culture supernatants can be detected in accurate amounts in the presence of other cytokines, RPMI media was spiked with rhIFN-γ and cytokines rhIL-12, rhIL-6, or rhTNFα, at concentrations of 30 and 400 pg/mL. There was no apparent interference by the cytokines in the detection of IFN-γ levels in the cell culture media (Table 13.7).

Table 13.7 IFN-γ detection is not affected in the presence of other cytokines

Cytokine[a]	IFN-γ (30 pg/mL) Expected[b]	IFN-γ (30 pg/mL) Observed[c]	% Recovery[d] of 30 pg/mL IFN-γ	IFN-γ (400 pg/mL) Expected[b]	IFN-γ (400 pg/mL) Observed[c]	% Recovery[d] of 400 pg/mL IFN-γ
None	33.77	26.15	77.44	407.28	512.49	125.83
rhIL-6	33.77	27.80	82.32	407.28	377.92	92.79
rhIL-12	33.77	28.22	83.55	407.28	382.57	93.93
rhTNF-alpha	33.77	27.62	81.79	407.28	392.45	96.36
Mean % Recoveries			81.27			102.23

[a] Cytokine concentrations identical to corresponding IFN-γ concentrations.
[b] Expected concentration is spiked plus 0.00 pg/mL background.
[c] Observed concentration is back-calculated concentration.
[d] Percent recovery is observed/expected x 100

Table 13.8 Matrix effects

Normal human sera dilutions	IFN-γ (30 pg/mL) Expected[a]	IFN-γ (30 pg/mL) Observed[b]	% Recovery[c] of 30 pg/mL IFN-γ	IFN-γ (400 pg/mL) Expected[a]	IFN-γ (400 pg/mL) Observed[b]	% Recovery[c] of 400 pg/mL IFN-γ
None	32.45	34.96	107.74	402.45	435.06	108.10
1:4	32.45	31.23	96.24	402.45	434.67	108.01
1:8	32.45	29.92	92.20	402.45	391.11	97.18
1:16	32.45	31.15	95.98	402.45	376.68	93.60

[a] Expected concentration is spiked plus 0.00 pg/mL background.
[b] Observed concentration is back-calculated concentration.
[c] Percent recovery is observed/expected 100.

Matrix effects

RPMI media was spiked with 30 or 400 pg/mL of rhIFN-γ and tested at final normal human sera dilutions of 1:4, 1:8, and 1:16. No interference of human sera on the detection of IFN-γ levels was noted at any of the dilutions tested (Table 13.8).

Stability

To determine the stability of the IFN-γ levels in cell culture supernatants subjected to freeze–thaw cycles, frozen aliquots were thawed and frozen on three different days, 2 days, or 1 day. The results (data not shown) indicate that IFN-γ levels remain stable and within ±25% of the freshly prepared sample in cell culture supernatants subjected to at least three freeze–thaw cycles.

Assay precision

Inter-assay variability was assessed using cell culture supernatants assayed in duplicate in three different assays on three separate days. The inter-assay coefficient of variation for three independent assays was 9.02%. Intra-assay variability was determined by assaying six replicates of cell culture supernatants in one assay. The pooled intra-assay coefficient of variation was 5.13%. Further, inter-operator variability was evaluated to be 1.75% between two trained individuals who performed two identical assays on the same day using cell culture supernatants. The inter-, intra- and inter-operator variability results confirm that the assay precision is high.

Validation of quality controls

A set of assay controls, Quality Control (QC), samples for IFN-γ ELISA were included on each IFN-γ ELISA plate. The Low, Mid, and High QCs were purchased from R&D Systems for the IFN-γ assay and within laboratory ranges were established. Acceptance levels were determined by taking the average of control results from three or more separate runs with ±3 standard deviation (SD). Further, the 1000 pg/mL standard, diluted 1:5 in assay buffer was substituted for one of the controls to mimic the dilution performed with test sample plates. Also, stability and assay precision was evaluated for QC samples similar to the procedures mentioned above for culture supernatants. The pooled estimate of the inter-assay coefficient of variation was 18.82% and intra-assay coefficient of variation was 3.67%. IFN-γ levels in QC samples remained stable after undergoing at least three freeze–thaw cycles and also remained stable for 1 month of storage at –70 °C.

13.6 Development and validation of IFN-γ induction assay

Development and validation of this NAb assay was performed in accordance with best practice approach as per ICH guidelines (ICH, 1997) along with other published reports (Gupta et al, 2007; Kelley et al, 2005; Mire-Sluis et al, 2004; Tacey et al, 2003; Wadhwa et al, 2003;1 Wei et al, 2004).

Assay development was initiated with optimization of cell culture procedures (media and cell seeding density). NK-92MI cells were obtained from ATCC (American Tissue Culture Corporation) and cell-seeding density was established from cell growth curves in different kinds of media. Cell culturing and response of the cells to induce IFN-γ in the presence of rhIL-12 was also compared in both RPMI based (used for NAb assay) and MEMα media recommended for cell maintenance. NK-92MI cells were cryopreserved at passage four following original thaw of cells from ATCC to create 15 Master Cell Bank vials from which a 120 vial Working NAb Cell Bank was generated. For optimum induction of IFN-γ responses in the NAb assay, cells should be maintained in MEMα media with a seeding density of 2×10^5 cells/mL and used in the assay at 48–72 hours following passage. Cells should have a density between 2×10^5 and 6×10^5 cells/mL and a viability >50% to be used in the induction assay conducted using RPMI media.

rhIL-12 dose response

Next, the optimum concentration of rhIL-12 was determined by measuring IFN-γ production at increasing concentrations of rhIL-12 (0.002 ng/mL to

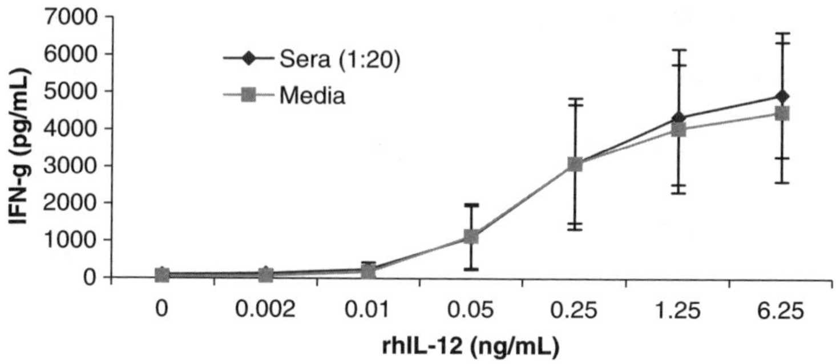

Figure 13.4 rhIL-12 dose response curve. Results represent mean of three experiments performed in both sera and media

6.25 ng/mL) in both sera and media-based matrices (Figure 13.4). There was close correlation of results between media and sera (5% NHS, reflecting 1:20 dilution of clinical sample) indicating minimal interference by assay matrix. Treatment of cells with 0.05 ng/mL rhIL-12 consistently showed a robust response of >70% increase of IFN-γ levels compared to treatment without rhIL-12 (cells alone) (Figure 13.4) with a good signal to noise ratio. Therefore 0.05 ng/mL concentration of rhIL-12 was selected for titrating the inhibition antibody control (DRUG Y) and positive NAb controls.

Dose response of inhibition antibody control

Inhibition of rhIL-12 induced IFN-γ levels at increasing concentrations of DRUG Y (1 ng/mL to 1 μg/mL) in both sera (5% normal human sera, NHS) and media-based matrices was tested. Increasing concentrations of DRUG Y up to 100 ng/mL proportionally inhibited rhIL-12 induced IFN-γ levels (Figure 13.5). Concentrations beyond 100 ng/mL did not show any further inhibition. There was a close correlation of results between media and sera with IC_{50} values of 10.26 ng/mL and 11.36 ng/mL, respectively. More than 50% of inhibition was observed at 25 ng/mL concentration of DRUG Y and was subsequently used to determine the optimum concentration of neutralization antibody control.

Dose response of positive and negative neutralization antibody controls

NAb positive controls are a valuable assay performance indicator. Reversal of DRUG Y inhibition of rhIL-12 induced INF-g levels was tested at increasing concentrations of DRUG Y NAb (anti-DRUG Y, at 1 ng/mL to 1 μg/mL) spiked into both sera (5% NHS) and media. In the sera-based matrix, more than 50% reversal of DRUG Y inhibition was observed at 100 ng/mL and complete reversal at 500 ng/mL concentration of monoclonal

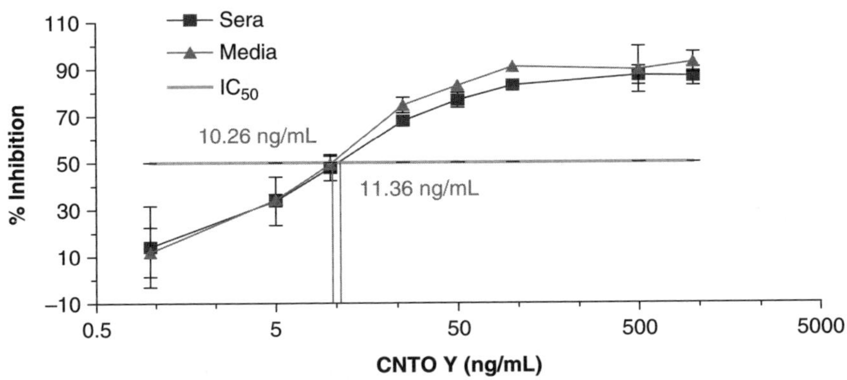

Figure 13.5 IFN-γ inhibitory response by DRUG Y. Shown are the dose response curves from mean of three experiments performed in both sera and media

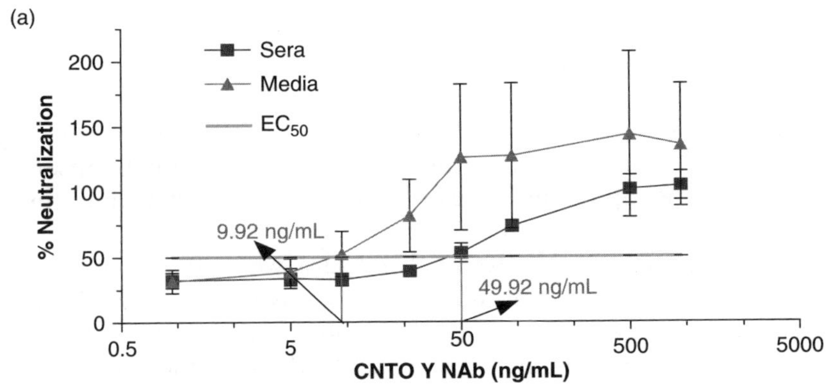

Figure 13.6 Reversal of DRUG Y inhibition of IFN-γ response by DRUG Y NAb (neutralizing murine monoclonal antibody). Shown are mean ± SD results from three experiments performed in both sera and media

NAb (Figure 13.6). The ED_{50} value for media was 9.92 ng/mL, whereas for sera (5%) it was 49.92 ng/mL. Since the NAb assay will be run with sera-based samples, monoclonal NAb concentration of 100 ng/mL was selected to use as NAb positive assay control.

Another NAb positive control, generated in Cynomolgus monkeys, included bleeds that tested positive in the EIA assay and were further tested in specificity and titration assays. Samples that tested positive in these assays were subsequently tested in the NAb bioassay at various dilutions (1:20 to 1:40 000) of the polyclonal sera. Figure 13.7 shows the dilutional NAb response of the pooled antisera from one monkey along with the lack of NAb response of the pre-bleed sera. Dilution

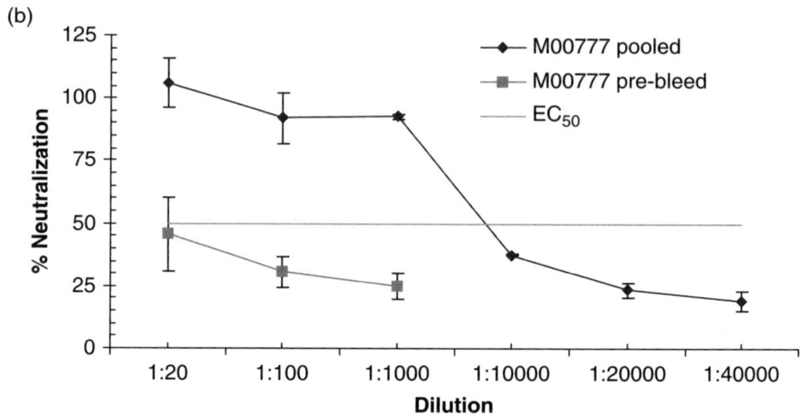

Figure 13.7 Dilutional response curve for pooled polyclonal monkey antisera to DRUG Y Reversal of DRUG Y inhibition of IFN-γ response by production bleed sera compared to no reversal of this response by pre-bleed sera. Data represent mean ± SD of three experiments performed in both sera and media

of 1:1000 and above showed more than 50% neutralization, with higher background at 1:20 ad 1:100 dilutions. Therefore, polyclonal antisera dilution of 1:1000 is optimum to use as positive neutralization response control.

Further, anti-DRUG Y polyclonal antisera were affinity purified and tested at increasing concentrations (1–500 ng/mL) for NAb activity. Increasing concentrations of the monkey polyclonal DRUG Y Abs (PAbs) up to 100 ng/mL proportionally neutralized DRUG Y reflected in increased rhIL-12 induced IFN-γ levels. More than 50% reversal of DRUG Y inhibition was observed at 100 ng/mL and complete reversal at 500 ng/mL. The EC_{50} values were 44.21 and 55.08 ng/mL for two PAbs and closely resemble anti-DRUG Y NAb results with murine monoclonal NAb. Any of these three NAb reagents can be used as a positive NAb control for clinical sample testing; however, polyclonal antisera is the most appropriate choice as it more closely represents clinical sample immune responses than other controls.

Interference of assay matrix

Sera samples from non-DRUG Y treated psoriasis ($n=8$) and MS ($n=8$) patients representative of clinical study subjects along with age and gender matched normal control sera were tested at different dilutions (1:10, 1:20, 1:40) for effects on IFN-γ production in the presence or absence of DRUG Y, along with background NAb response. Similar background

NAb response (without non-specific stimulatory effect) and percent inhibition was observed at all dilutions, however, higher IFN-γ responses were observed (equal to media) with good signal to noise ratio (four fold response over background) at 1:20 and 1:40 dilutions in normal and disease matrices. Therefore clinical samples would be tested at 1:20 dilution of the sera.

Further, Sample Controls (SC1 and SC2) included in the NAb assay would serve as the specificity controls for the sample and would indicate any stimulatory or inhibitory effect of the sample itself. Also pre-treatment sample from the same patient as the post-treatment sample is often the best control to exclude any non-specific effects.

Assay specificity

Specificity was addressed by assaying serum samples spiked with non-specific Abs and NAbs, along with cytokines such as IL-6, TNF-α, and IL-2. Results indicated minimal interference of therapeutic Abs or other anti-idiotypic Abs (data not shown) or other cytokines. However, TNF-α showed increased background (due to possible stimulatory effect of TNF-α by itself on NK cells to produce IFN-γ, though much less than IL-12 induction) but did not affect the ability of the assay to detect anti-DRUG Y neutralizing antibodies. Also, physiological serum levels of TNF-α in clinical samples are anticipated to be much lower than the concentration of TNF-α used in these experiments.

Assay cut-off

The NAb assay cut-off for potentially positive samples was determined based on the baseline response from treatment naïve serum samples from normal (18), psoriasis (42) and (21) multiple sclerosis subjects. Statistical analyses of 141 data points (percent IFN-γ recovery results, TS) were used to determine the NAb assay cut-off. Percent recovery is defined as (TS-SC1)/(SC2-SC1)×100. In addition, 96 results from 3 NHS pools on 71 plates were included to help estimate the variance components. Also, 77 plates from 34 assay dates that included 9 DRUG Y thaw dates were considered to determine variance. Apart from these significant variance components, other non-significant factors (cell passage number, cell thaw number, DRUG Y lots, freeze–thaw of sample and critical reagents) were also considered.

$$
\begin{aligned}
&\text{The primary analyses are based on } 100 \times \ln\ (\%\ \text{recovery}) \\
&\quad = 100 \times \ln\ [(\text{TS–SC1})/(\text{SC2–SC1}) \times 100] \\
&\quad = 100 \times \ln\ (\text{TS–SC1}) - 100 \times \ln\ (\text{SC2–SC1}) + 100 \times \ln\ (100)
\end{aligned}
$$

where

TS = Test condition IFN-γ of {cells + 1:20 dilution of sample + rhIL-12 + DRUG Y}
SC2 = Maximum IFN-γ of {cells + 1:20 dilution of sample + rhIL-12}
SC1 = Background IFN-γ of {cells + 1:20 dilution of sample}
ln = natural logarithm

The overall geometric mean percent IFN-γ recovery is 20.5 (3.0186 in the ln scale) with approximate total inter-assay CV of a naïve donor's percent IFN-γ recovery of 26.89% (0.2689 SD in the ln scale). The assay cut off was determined to be greater than or equal to 36% including 2.05 standard deviations above the mean and a 2.0% naïve sample positive rate.

Assay sensitivity and drug interference

The lowest concentration of positive NAb controls that showed percent neutralization greater than or equal to 36% was determined by spiking two samples of 5% NHS with affinity purified monkey DRUG Y PAbs (M00777 or M00778). Results indicate that the limit of detection of NAbs for the assay was estimated as low as 552.80 to 756.80 ng/mL in 100% serum (20×) based on purified monkey polyclonal antibodies (PAbs) using the 36% assay cut off value representing 2% false positive rate.

Clinical samples from DRUG Y dosed subjects may have residual drug remaining in the sera depending on the time point for IR analysis. In order to test the effect of circulating serum drug concentration on the NAb assay, mock samples were created using DRUG Y spiked in sera at various concentrations at 20× to attain final concentration ranging from 0.0001 to 5000 ng/mL in 5% serum matrix. This is in addition to the normal assay level (25 ng/mL) of DRUG Y used along with 100 ng/mL of either DRUG Y NAb (at 20× dilution), or purified polyclonal monkey anti-DRUG Y PAb (at 40× dilution) or 1:1000 dilution of polyclonal monkey serum (at 40× dilution).

In the presence of DRUG Y NAb (mAb) and purified monkey anti-DRUG Y PAb, the NAb assay could tolerate 0.070 and 1.408 μg/mL of residual DRUG Y in 100% serum, respectively. With polyclonal monkey serum, the assay could tolerate much higher levels (up to 4 μg/mL) of residual DRUG Y in 100% serum. However the true sensitivity and drug interference should be evaluated from pooled clinical samples that indicate high NAb activity.

Assay precision

Intra-assay variability was determined by assaying six replicates of cell culture supernatants from assay controls in one assay. The intra-assay %CV for all conditions (with or without DRUG Y NAb) ranged from 5.16 to 13.23. The intra-assay %CV was 7.78 for polyclonal monkey sera and 4.32 for purified polyclonal monkey anti-DRUG Y PAb. The pooled intra-assay %CV for all assay controls was 8.30 indicating acceptable assay precision within a plate.

Inter-assay variability was assessed using cell culture supernatants from assay controls generated in three different plates on the same day using cells from a single culture flask. The inter-assay %CV for the three assay plates ranged from 3.95 to 20.05 and was 2.04 for polyclonal monkey sera. The pooled inter-assay %CV for all conditions was 9.59. The polyclonal anti-DRUG Y serum functioned in a reliable manner in its ability to reverse DRUG Y inhibitory activity and is suitable to be included as a positive NAb control for clinical sample testing.

Inter-assay variability

Inter-assay variability was assessed using cell culture supernatants from assays set up on 10 different days. Results indicated that inter-day variability of NAb assay was much higher than within-plate and within-day variability, with a pooled %CV for all assay controls of 33.60%. Part of the inter-day variability is also contributed by variation in the cell density and viability. Therefore, testing of all related samples (e.g. all time points of a patient) should be performed on the same day.

Inter-operator variability

Inter-operator variability was evaluated between three trained operators by performing two identical assays on the same day. The assessment of inter-operator variability showed a pooled %CV range of 5.05 to 10.05. These results confirm the precision of the NAb assay procedure when performed by trained operators.

13.7 NAb bioassay robustness

Effect of cell passage

To determine the effect of the cell passage number on their response to rhIL-12, cells were cryopreserved, thawed, and assayed together at passages 4, 10, 15, 22, 28, and 35. Results showed that NK-92MI cells showed comparable IFN-γ responses up to passage 35 with minimum inter-passage variability (Table 13.9). Therefore, clinical samples will be tested in the NAb assay using cells until passage 35 after which a fresh culture will be started for any further analysis.

Table 13.9 Effect of passage number

Passage number	% Inhibition with DRUG Y	% Recovery of IFN-γ with NAb, DRUG 1438
4	68.25	82.27
10	63.88	87.69
15	73.15	91.80
22	74.99	88.92
28	76.52	95.82
35	68.57	105.63

Effect of NAb assay culture time

To determine the effects of culture time of 20 ± 4 hours on IFN-γ responses, supernatants were collected from NAb assay plate at 16, 18, 20, 22, and 24 hours of culture and IFN-γ levels were measured. Results indicated that culture time of 20 ± 2 hours do not increase background response and showed similar percent inhibition and percent recovery of IFN-γ response, with a %CV range of 2.46 to 16.97. However, all related samples must be tested using the same culture time.

Lot-to-lot variability of critical reagents

Over the course of the NAb assay validation, it was necessary to use different lots of critical reagents either due to reagent expiration or consumption. In order to evaluate the lot-to-lot variability of these reagents, two separate lots each of rhIL-12, DRUG Y, and DRUG Y NAb were tested in the same assay under identical conditions. Variability between different reagent lots ranged from 3.87 to 20.31 %CV indicating that different lots of critical reagents may be used during a study, though samples from all time points of a patient must be tested using the same lot of reagents.

Stability of assay supernatants

Stability of the IFN-γ levels in assay supernatants subjected to three cycles of freeze–thaw was determined using assay controls. The results indicated that IFN-γ levels remained stable with a %CV range of 0.86 to 10.14 for fresh and frozen supernatants subjected to at least three freeze–thaw cycles. More than three freeze–thaw cycles are not recommended.

Stability of assay supernatants stored at 4 °C or –70 °C for 2 to 8 weeks was examined and results indicated that IFN-γ levels remained stable after long-term storage up to 4 weeks at both 4 and –70 °C and were comparable to results obtained with samples stored overnight.

Stability of functional response of stored assay control reagents

Spiked assay controls may need to be stored frozen and used for multiple assays. Response from M00777 sera was stable up to 4 weeks of storage at 4 °C. Frozen DRUG Y and NAb controls showed stable responses for up to 8 weeks of storage at –70 °C, while the M00777 anti-DRUG Y sera was stable for up to 4 weeks.

13.8 Result output, interpretation, assay/plate and data acceptance criteria for clinical samples

The result output obtained from the NAb assay is pg/mL of IFN-γ levels. If the IFN-γ ELISA plate passes based on the performance of IFN-γ standards and quality controls, further evaluation for accepting NAb assay sample results will be performed as per the following criteria.

Acceptance criteria for standards and QCs of IFN-γ ELISA

- At least 10 out of the 14 standard curve calibrator replicates must be within ±25% of their nominal concentration with %CV of replicate OD values to be ≤25%. The back-calculated concentration of at least one replicate per standard must be within ±25% of its respective nominal concentration.
- At least four of the six back-calculated QC sample means on each plate must fall within the laboratory-determined range (the mean of three or more results ±3 standard deviations, determined for each new lot of controls).
- For the 1:5 diluted 1000 pg/mL standard, the mean result must be within ±25% of its nominal concentration.
- The %CV for the two replicates of the QCs with a mean value within the accepted range must be ≤25%.

Acceptance criteria for NAb bioassay control samples

NAb assay acceptance will be examined once the IFN-γ assay results are accepted based on the performance of the standards and controls and sample replicates as outlined above.
The following criteria must be met for the NAb assay plate to be accepted:

- The %CV of the back-calculated assay control result replicates must be ≤30%.
- The OD of each NAb assay control sample replicate must be within the interpolated range of mean OD values for 15.60 pg/mL (LLOQ) and

1000.00 pg/mL (ULOQ), except for the background response control (AC1) and the maximum inhibitory (DRUG Y) response control (AC3).

- The mean of the maximum IFN-γ response control (AC2) replicates should be four fold over the mean of the background response control (AC1) replicates (AC2/AC1≥4).
- The mean of the maximum IFN-γ response control (AC2) replicates should be inhibited by the mean of the maximum inhibitory (DRUG Y) response control (AC3) replicates by more than two fold (>50%) (AC2/AC3≥2).
- The mean of the maximum response (AC2) replicates should be less than two fold over the mean of the positive NAb control (AC2/AC4≤2). Less than 50% inhibition of maximum response suggests more than 50% neutralization potential of the positive NAb control.

Clinical sample result acceptance criteria

- The %CV of the back-calculated test sample result replicates must be ≤30%.
- The mean of the maximum sample response (SC2) replicates should be four fold over the mean of the sample background (SC1) replicates (SC2/SC1≥4).

Clinical sample result output for presence of NAbs against DRUG Y

NAb assay results for clinical samples are expressed as percent recovery of IFN-γ response = ((TS-SC1)/(SC2-SC1)) × 100. Test sample results of <36% will be reported as NAb negative. Any results of ≥36% (assay cut-off) are interpreted as positive for presence of anti-DRUG Y NAbs.

13.9 Summarizing remarks

In summary, validation of two functional cell-based Nab assays has been described. While the two assays have different endpoint measurements, the parameters evaluated in both assays overlap to a large extent. The performance of the target cell in culture is one of the most important aspects that needs to be evaluated to ensure consistent and reproducible results every time a clinical sample is evaluated. We have successfully used both these assays to evaluate the presence of neutralizing Nabs in samples from clinical trials.

References

Casadevall N, Nataf J, Viron B et al (2002). Pure red cell aplasia associated with anti-erythropoetin antibodies in patients treated with recombinant erythropoietin. *N Engl J Med*, **109**, 71–78.

DeSilva B, Smith W, Weiner R, Kelley M, Smolec J, Lee B, Khan M, Tacey R, Hill H and Celniker A. (2003). Recommendations for the bioanalytical method validation of ligand-binding assays to support pharmacokinetic assessments of macromolecules. *Pharm Res*, **20**, 1885–900.

Findlay JW, Smith WC, Lee JW, Nordblom GD, Das I, DeSilva BS, Khan MN and Bowsher RR (2000). Validation of immunoassays for bioanalysis: a pharmaceutical industry perspective. *J Pharm Biomed Anal*, **21**, 1249–1273.

Geng D, Shankar G, Schantz, A, Rajadhyaksha M, Davis H and Wagner, C (2005). Validation of immunoassays used to assess immunogenicity to therapeutic monoclonal antibodies. *J Pharm Biomed Anal*, **39**, 364–75.

Gupta S, Indelicato SR, Jethwa V, Kawabata T, Kelley M, Mire-Sluis AR, Richards SM, Rup B, Shores E, Swanson SJ (2007). Recommendations for the design, optimization, and qualification of cell-based assays used for the detection of neutralizing antibody responses elicited to biological therapeutics. *J Immunol Methods* **321**, 1–18.

International Conference on Harmonization (ICH) of Technical Requirements for Registration of Pharmaceuticals for Human Use (1997). Section Q2B: Validation of Analytical Procedures: Methodology.

Kelley M, Cooper C, Matticoli A and Greway A (2005). The detection of anti-erythropoietin antibodies in human serum and plasma: part II. Validation of a semi-quantitative 3H-thymidine uptake assay for neutralizing antibodies. *J Immunol Methods*, **300**, 179–191.

Mire-Sluis AR. (2001). Progress in the use of biological assays during the development of Biotechnology products. *Pharm Res*, **18**, 1239–1246.

Mire-Sluis AR, Barrett YC, Devanarayan V, Koren E, Liu H, Maia M, Parish T, Scott G, Shankar G, Shores E, Swanson SJ, Taniguchi G, Wierda D and Zuckerman LA (2004). Recommendations for the design and optimization of immunoassays used in the detection of host antibodies against biotechnology products. *J Immunol Methods*, **289**, 1–16.

Rosenberg AS and Worobec AS (2004). *Biopharm Int*, **17**, 12. Sciences Module, pp 34–42.

Schellekens H (2002). *Clin Ther*, **24**, 1720–1740.

Tacey R, Greway A, Smiell J, Power D, Kromminga A, Daha M, Casadevall N and Kelley M (2003). The detection of anti-erythropoietin antibodies in human serum and plasma. Part I. Validation of the protocol for a radioimmunoprecipitation assay. *J Immunol Methods*, **283**, 317–329.

Wadhwa M, Bird C, Dilger P, Gaines-Das R and Thorpe R (2003). Strategies for detection, measurement and characterization of unwanted antibodies induced by therapeutic biologicals. *J Immunol Methods*, **278**, 1–17.

Wei X, Swanson SJ and Gupta S (2004). Development and validation of a cell-based bioassay for the detection of neutralizing antibodies against recombinant human erythropoietin in clinical studies. *J Immunol Methods*, **293**, 115–126.

14 Endpoint assays in HIV-1 vaccine trials: functioning in a Good Laboratory Practices environment

Patricia D'Souza[1], Josephine H. Cox[2], Guido Ferrari[3], Nina Thapa Kunwar[1], Victoria Polonis[2] and Marcella Sarzotti-Kelsoe[3]

[1]*Vaccine Clinical Research Branch, Division of AIDS, NIAID, NIH, 6700-B Rockledge Drive – MSC 7628, Bethesda, MD 20892-7628, USA*

[2]*Walter Reed Army Institute of Research, US Military HIV-1 Research Program, Suite 200, 13 Taft Court, Rockville, MD 20850, USA*

[3]*Department of Experimental Surgery, Duke University Medical Center, P.O. Box 2926, Durham, NC 27710, USA*

14.1 Introduction

The goal of human immunodeficiency virus (HIV-1) vaccine development is to provide a safe means for individuals to establish durable and potent immunologic memory that can protect against HIV-1 infection.

Upon HIV-1 exposure, the vaccinated host should mount a rapid effector immune response that can destroy free and cell-associated virus. Much effort has been devoted to the development of an HIV-1 vaccine: to date, over 90 trials have been conducted with more than 50 different vaccines, yet success remains elusive and several obstacles remain towards this goal. These include the extreme diversity of the virus (Gaschen et al, 2002), even within a single person, the evidence that HIV-1 infects virus-specific CD4+ T cells (Douek et al, 2002) which are critical for maintenance of effective immunity in animal models of chronic infection, and the fact that the virus can integrate into the host genome and remain in a latent phase during which it is not recognized by adaptive immune responses (Blankson et al, 2002).

Despite the obstacles, there is evidence to support the feasibility of a vaccine for HIV-1. Rare instances of apparent protection from infection in highly exposed, persistently seronegatitve persons exist, and reports of adaptive immune responses suggest that protection from infection might occur (Kaul et al, 2001; Kebba et al, 2004; Lopalco et al, 2005). There are other reports where infection has occurred, but where the virus appears to be effectively controlled, suggesting that long- and short-term protection from disease progression, rather than protection from infection can occur (Zhu et al, 2003). These include the 1000-fold or greater drop in peak viremia in acute infection in most infected persons, and the small subset of infected persons who have low to undetectable viral loads, some as many as 25 years after initial infection (Buchbinder and Vittinghoff, 1999; Addo et al, 2003; Rosenberg et al, 1997). Incorporating these yet undefined elements of immunity that impart antiviral effect into a vaccine would be highly desirable.

A successful vaccine will likely need to elicit CD8+ T cell responses to HIV-1, together with virus-specific CD4+ T cells and neutralizing antibodies to protect from infection or from disease progression. The exact nature of the vaccine-induced memory response that will afford protection from HIV-1 infection or limit its spread is unknown. An effective B cell response to HIV-1 infection should elicit cross-reactive antibody against multiple HIV-1 strains, directed against an array of HIV-1 neutralizing epitopes, comprised of high titer antibodies, and maintained long term. In contrast, the required distribution, functional potential, and phenotype of a protective CD8+ T cell response to HIV-1 are unknown. The general consensus is that a successful vaccine should generate a large CD8+ T cell response to a breadth of cross-reactive HIV-1 epitopes that is qualitatively better than the CD8+ T cell responses elicited during natural infection.

To advance HIV-1 vaccine development it is essential to comprehensively define the immune responses that candidate vaccines elicit by employing technologies that are validated and robust so as to accurately prioritize and advance vaccine candidates based on immunologic potency. These types of datasets can help to define the components of vaccine induced immunity to protect against HIV-1 infection and assist in the design of efficacy trials to uncover mechanisms and predictors of protection.

In this chapter, the major focus will be on describing the elements of a laboratory program that will lead to the development of a safe and efficacious HIV-1 vaccine for humans; a process of peripheral blood mononuclear cell (PBMC) collection, processing, storage and analyses; a network of site-affiliated laboratories to conduct safety tests; a central laboratory to perform validated endpoint assays to help identify immune responses that correlate with vaccine-induced protection in Phase IIB to III efficacy trials; and a rigorous quality control and control assurance program for the entire operation.

14.2 Key components of a laboratory program

Specimen processing

Multiple parameters during the collection, processing, and storage of cryopreserved PBMCs can impact the performance of those cells during immune monitoring in clinical trials. Because early product advancement decisions are based on the data obtained from cellular immune assays, the quality of the cell samples obtained at the time of processing and during the storage will influence their biological function upon *in vitro* stimulation with the appropriate antigens. It may also influence the level of background obtained in the unstimulated/control conditions (see Table 14.1).

Recently, the HIV-1 Vaccine Trials Network (HVTN) investigated the effects of a variety of different parameters on recovery, viability, and function of cryopreserved T cells. These parameters included: (1) three different anticoagulants (sodium heparin (Na-Hep), acid citrate dextrose (ACD), and ethylenediamine tetraacetic acid (EDTA); (2) two different times of processing from collection (6–8 versus 24 hours); (3) two different separation methods (Accuspin® tubes or cell preparation tubes (Vacutainer® CPT™); (4) two different shipping conditions for fresh samples (whole blood in collection tubes versus separated PBMC in cold medium); and (5) two different freezing media. Data from these studies show a decrease in cell viability after thawing samples that were shipped overnight using EDTA and Na-Hep when compared to the same samples isolated within 8 hours, and to samples collected in ACD. Recoveries and viabilities were significantly lower for ACD when the 8 hours processed samples were

Table 14.1 Peripheral blood mononuclear cells processing

Step	Factors	Parameters to be evaluated[a]
1. Collection	Anticoagulant	Cell yield, recovery and viability
2. Separation	Time from venipuncture Method of separation	Cell yield, recovery and viability
3. Cryopreservation		
A. Freezing medium	Type of serum % DMSO	Effect on long-term storage of cell recovery, viability and biological function
B. Cryopreservation method	Controlled rate freezer Mr. Frosty device	Effect on long-term storage on cell recovery, viability and biological function

[a] The parameters should be evaluated at time of separation and after thawing cryopreserved samples.

compared to the 24 hours-processed after thawing and overnight incubation. However, there were no significant differences in the following parameters: (1) the two different separation methods; (2) the shipping condition of cells prior to cryopreservation; (3) the freezing medium of choice. The data on recovery and viability reflect the biological function of the cryopreserved samples using both the interferon gamma (IFN-γ) ELISpot and intracellular cytokine (ICS) assays. Overall, the authors suggest that the length of time from venipuncture to cryopreservation is the single most important parameter for determining performance of T cells in cellular immune assays and that optimally blood samples should be processed within 6–8 hours after collection (Bull et al, unpublished work).

These data imply that clinical trial sites associated with testing HIV-1 vaccines must be proficient at PBMC isolation and cryopreservation, or be close to contract laboratories that can provide this service. If the clinical trial involves multiple clinical sites, on site sample storage and shipping conditions to the repository also require optimization as any of these steps can influence subsequent steps and it is unrealistic to predict what combination will work best without investigating different conditions. Moreover, a quality assurance/quality control program aimed at evaluating the competency of each operator/laboratory processing blood should occur, along with a program to ensure uniform processing of specimens over time within the testing laboratory.

14.3 Clinical diagnostic and endpoint tests

Safety testing

The safety of HIV-1 candidate vaccines is the primary endpoint in Phase I and II vaccine trials. Consequently, a number of clinical tests are performed routinely to screen and monitor trial participants. Initial screening includes historical and clinical evaluation for indicators of compromised health such as liver disease, hypertension, autoimmune disorders, and sexually transmitted diseases. Additional clinical laboratory testing is performed for blood chemistries, complete blood count (CBC), pregnancy tests in female volunteers, and other functional tests such as urinalysis and tests to detect sexually transmitted diseases. These types of abnormalities may complicate the interpretation of the vaccine's safety profile. During the conduct of the trial, similar laboratory tests are also performed to evaluate whether administration of the study product has had an adverse effect on the individual.

Testing for HIV-1 infection

Immune responses to candidate HIV-1 vaccines have the potential to be confused with responses to actual HIV-1 infection by standard serological diagnostic testing because vaccine-induced antibodies may cause specimens to be categorized as positive by some HIV-1 screening tests. Consequently, testing for HIV-1 infection is an essential component of the battery of safety tests in HIV-1 clinical trials. Elaborate testing algorithms have been instituted world-wide to discriminate HIV-1 infected individuals from non-infected individuals as well as uninfected, vaccinated individuals.

Enzyme-linked immunosorbent assays (EIA)

Specimens are initially assayed with an EIA test. Negative test results indicate that the participant is not infected with HIV-1. Positive EIA results are repeated in duplicate. If either of the duplicate tests is reactive, the serum sample is tested by Western blot. Positive HIV-1 Western blot test patterns are further confirmed by HIV-1 nucleic acid tests (NAT) using different formats of polymerase chain reaction (PCR).

The EIA is a 1 day test that allows for inexpensive, high throughput screening for HIV-1 antibodies. An HIV-1 EIA is usually performed using serum from blood; most EIA methods follow a similar format. Briefly, inactivated HIV-1 antigens are pre-coated onto an ELISA plate, dilutions of test sera are added, and positive and negative reactivity controls are also added to separate wells. Antibodies in seropositive serum will bind to the HIV-1 antigens, after which a secondary anti-human immunoglobulin (Ig) preparation coupled to an enzyme, such as horseradish peroxidase

(HRP), is added to the wells. After washing, a substrate, such as chromogen, which changes color when cleaved by the enzyme attached to the secondary antibody, is added to the wells. High levels of anti-HIV-1 antibodies bound to the HIV-1 antigens in the plate correlate with an increase in enzymatic cleavage of substrate and therefore, more color development. The level of colored precipitant formed is quantified by measuring the optical density of the fluid in each well using a plate reader (Aldovini and Walker, 1990).

In laboratories that perform clinical testing, there are numerous quality control (QC) samples built into a validated assay; the QC samples must meet certain pre-assigned criteria for the test serum assay results to be accepted. If the QC samples on a given EIA plate fail, then the entire plate must be rejected and those samples retested. The cut-off values indicating a positive or negative result are statistically pre-determined during the validation phase and are used at the data analysis phase to determine the reportable value for a sample (Pombo et al, 2004).

Western blots

The main confirmatory test of choice for serologic diagnosis of HIV-1 infection in most laboratories is the Western blot or standard immunoblot. For this test, a preparation of HIV-1 antigens (typically from a cell-free viral lysate) is separated by size on a sodium dodecyl sulfate poly-acrylamide gel (SDS-PAGE), and these separated antigens are subsequently transferred to a solid matrix such as nitrocellulose paper. Strips of the nitrocellulose can then be reacted with sera and any anti-HIV-1 antibodies will bind to the specific antigens. After incubation of the strips with an enzyme-conjugated secondary antibody (anti-human IgG) or biotinylated secondary antibody followed by enzyme-conjugated avidin or streptavidin, strips are incubated with substrate.

A characteristic pattern of bands may then be visualized on the strip, indicating the presence of anti-HIV-1 antibodies. The three major structural genes in HIV-1, envelope (env), polymerase (pol) and group-specific antigens (gag), give rise to nine major viral gene products that may be visualized on an immunoblot. The nine products derived from the three gene groups are as follows: env gives rise to gp160, gp120 and gp41; pol generates p65 reverse transcriptase (RT), p53 RT, and p34 endonuclease; and gag proteins comprised of the p55 precursor, p24 core, and p17 matrix or amino-terminal gag (Travers and Kanki, 1990). Currently, for the two US FDA approved Western blot kits available on the market, there is considered to be three band groups: gp160/gp120; gp41; and p24. The acceptance criteria for an HIV-1-positive Western blot test for many laboratories, is that any two of these three band groups must be reactive on the immunoblot strip (Personal communication (2006), COL Nelson Michael, MD, Director, The

US Military HIV Research Program and The Division of Retrovirology, Walter Reed Army Institute of Research, 1600 Gude Drive, Rockville, MD 20850).

The availability of these commercial kits has eliminated the need for immunoblot assay validation in most laboratories, but the interpretation of the results from the QC samples included in these test kits must be strictly adhered to for these kits to be used for diagnostic purposes. If the Western blot is negative, the participant is diagnosed as "Uninfected HIV-1". If the Western blot is indeterminate or positive, then the algorithm will be continued through the NAT. The algorithm may be repeated with newly collected samples, if necessary, until an infection status determination can be made. If the Western blot and NAT are positive, the sample is considered presumptive HIV-1 infected and a second, different, specimen is requested for confirmation. To diagnose HIV-1 infection, the laboratory must obtain an evolving (i.e., gain bands) positive HIV-1 Western blot and HIV-1 NAT positive results on two different specimens. Some specimens from participants receiving HIV-1 vaccines with envelope boost show 'positive' Western blot patterns (gp160/120, gp41, p24). These participant patterns do not evolve over time. A full assessment may come only over time, with multiple blood draws; uninfected participants will have Western blot patterns that do not evolve.

An individual not infected with HIV-1 may have cross-reactive antibodies to an HIV-1 protein such as gag p24, resulting in a false positive HIV-1 EIA. In addition, women who have had multiple pregnancies may possess antibodies directed against human leukocyte antigens (HLA) which are present on the host cells used to propagate HIV-1. As HIV-1 buds from the surface of the host cell, it incorporates some of the host cell HLA into its envelope; if these HLA molecules are present in the viral antigens used to coat the EIA wells, the antibodies to HLA will bind to the HLA in the wells, causing a false positive result. Certain people may also have a substance in the blood that interferes with proper reactivity in the EIA, or may be poor producers of antibody. The amount of antibody, consequently, may be too low to measure accurately or may go undetected. This result is termed a false negative. False negatives can occur during the window between infection and an antibody response to the virus (seroconversion). Thus, while EIA for HIV-1 inexpensive and high throughput, the limitations of the assay may be false negatives, false positives, and a need for confirmatory testing.

Within the context of an HIV-1 vaccine trial, HIV-1 diagnostic testing must be performed in a manner that retains the blinding of the participant's vaccination status. To preserve blinding, the vaccine clinic receives a summary interpretation that the participant is "Infected HIV-1" or "Uninfected HIV-1". This avoids disclosure of HIV-1 antibody status as a result of participating in an HIV-1 vaccine trial, while providing the participant his or her HIV-1 infection status. When an indeterminate or a presumptive positive result is obtained, further repeat testing is required on a new specimen. At the end of the study, the participants are unblinded and advised about future testing outside of the vaccine clinic setting.

Immune responses elicited in individuals immunized with increasingly complex HIV-1 vaccines are more likely to yield positive EIA and Western Blot reactivities. Vaccinated individuals who are EIA and/or Western blot positive may have problems when applying for jobs, obtaining health care provision or trying to pass immigration requirements where a blood screen is required. Participants in HIV-1 vaccine trials are provided with the means to indicate that they have participated in an HIV-1 vaccine trial, but this can still be problematic (Allen et al, 2001; Belshe et al, 1994). Moreover, there is a stigma associated with HIV-1 seropositivity; a vaccinated individual may face problems with family and community members who don't understand the complexities of HIV-1 vaccine trials, and assume that the volunteer is trying to conceal an HIV-1 infection status.

Thus, there is now a concerted effort to design diagnostic tests that can distinguish vaccinated from HIV-1 infected volunteers. A recently published study offers the hope of a simple and sensitive EIA to accomplish this goal (Khurana et al, 2006). This paper used a whole-HIV-1-genome phage display library to identify conserved sequences in Env-gp41 and Gag-p6 which are recognized soon after infection, do not contain protective epitopes, and are not part of most current HIV-1 vaccines. An EIA assay was recently developed (HIV SELECTEST) which could not only distinguish vaccinated from HIV-1 infected individuals with >99% specificity and sensitivity, but could also detect HIV-1 infection earlier than standard EIA assays. Additional testing of the HIV SELECTEST is planned within the context of HIV-1 vaccine efficacy trials with clinical specimens from different clades, and using specimens from non-HIV-1 infections such as flu. If this type of sophisticated test fails to distinguish vaccinated from HIV-1 positive samples, then it will be necessary to rely increasingly on NAT testing in conjunction with serological testing or NAT exclusively (Pilcher et al, 2005).

Endpoint assessments

Defining HIV-1 vaccine efficacy and correlates of protection from HIV-1 disease is difficult since there are no early clinical markers of infection other than a rapidly resolving fever. Typically clinical manifestations do not appear until months or years after HIV-1 infection (Mindel and Tenant-Flowers, 2001). Thus, for HIV-1 infection there was a need to define surrogate markers of infection. Early in the epidemic, two markers were used to track the course of the disease; namely antibodies to HIV-1 Gag and or Env and the number of T cells expressing CD4 in the blood. Antibody responses are not usually reliably detected until six or more weeks after infection (seroconversion) although new technologies are being used to detect antibody responses earlier (Khurana et al, 2006). The EIA assay for detection of p24 antigen became the cornerstone of public health initiatives for keeping blood supplies safe for transfusion and other applications.

Antibody based tests are not good markers of acute infection or for tracking the chronic phase of the disease. CD4+ T cell counts can be used to track disease course but again are not very useful for tracking acute HIV-1 disease. As recombinant DNA technologies have improved, the measurement of viral load by PCR technology became one of the key markers for disease staging. In general there are very well described clinical and surrogate laboratory markers that can be used to track disease progression, drug and other therapeutic interventions (Kiepiela et al, 2005). Refinements in PCR technology have occurred systematically and frequently resulting in assays that can be used for multiple clades and to detect early infection (Jagodzinski et al, 2003; Pilcher et al, 2005).

For measuring endpoints in HIV-1 vaccine trials, the primary endpoint has to be acquisition of HIV-1 infection and this must be measured as soon as feasible to capture incident infection, curb transmission, and utilize appropriate patient management strategies. In small phase I and II trials in which typically low-risk volunteers are enrolled, frequent safety and immunogenicity testing is performed. In efficacy trials, in at-risk populations, the timing of visits to determine vaccine efficacy is typically every 6 months. Thus, persons who become infected between clinic visits would most likely seroconvert and appear positive by EIA and Western blot technologies.

Current vaccine trials are testing candidates that are able to generate immune responses in animal models which control viral replication and prevent disease following virus challenge, rather than preventing HIV-1 acquisition. In humans, the set point viral load predicts the subsequent disease course (Mellors et al, 1997), whereas HIV-1 transmission risk is strongly associated with peak viral load (Quinn et al, 2000). Recent data in the simian immunodeficiency virus (SIV)/macaque animal model suggest that it is the peak viral load, rather than set-point viral RNA levels, as well as preservation of memory CD4+ T cells that are predictors of vaccine efficacy and correlates of protection from disease, as conferred by cellular immune-based vaccines (Letvin et al, 2006). As new HIV-1 vaccine efficacy trials are being conducted and planned, more sophisticated endpoint assessments are being considered. For example, new trial designs are evaluating whether an HIV-1 vaccine can reduce the spike in HIV-1 viremia that is associated with acute infection relative to levels among placebo recipients who become infected. If viral load measures are taken early after peak viremia, but before initiation of antiretroviral therapy, which is now becoming globally accessible, measuring viral load is unlikely to be confounded by early initiation of antiretroviral therapy. Early measures of viral load could also provide important surrogate information on the potential impact of vaccination on HIV-1 transmission, as peak viral load correlates with risk of HIV-1 transmission in serodiscordant partner studies (Gray et al, 2001). Efforts are also underway to quantitate the central memory CD4 + T cells in vaccinated and infected individuals as this marker may be an important immune correlate of long-term protection and predict HIV-1 vaccine efficacy.

To date two phase III HIV-1 vaccine efficacy trials have been conducted for which the primary endpoint was acquisition of HIV-1 infection as measured by detection of HIV-1 antibodies, using standard HIV-1 EIA and Western blot assays. Secondary endpoints included viral loads, CD4 counts, rates of antiretroviral-therapy initiation, and the genetic characteristics of the infecting HIV-1 strains between treatment arms. To assess the impact of vaccination on viral load, accurate measurements are critical. Half log or more reductions in viral load result in slower progression of HIV-1 disease and improved survival. Table 14.2 shows tests typically used to determine HIV-1 infection status and vaccine efficacy and caveats of their use in HIV-1 vaccine settings. In addition to the advantages and disadvantages listed in Table 14.2, other practical considerations need to be taken into account; (1) can all laboratory assays be performed on site or will some highly technical assays require centralization, (2) laboratory capacity for specimen processing, storage and shipping, (3) turnaround time for endpoint assays, and (4) cost to perform one or more endpoint assays.

14.4 Vaccine pipeline and clinical trials

A full discussion of the vaccine pipeline is beyond the scope of this article and the reader is referred to some excellent resources and reviews (Garber et al, 2004; Graham, 2002; Letvin, 2005; McMichael, 2006; Spearman, 2003). Recent vaccine candidates advancing in clinical trials have focused primarily on inducing cellular immunity using gene-based vectors, such as DNA and replication defective adenoviruses or poxvirus. These vaccines primarily intend to stimulate either the CD4+ and CD8+ T cell responses or CD8+ T cells alone. The interplay between these cells influences the strength and duration of the CTL response. With the emphasis on vaccines that can induce T cell mediated immune responses, surrogate endpoints such as the ability of effector cells to secrete multiple cytokines are under consideration (Pantaleo and Harari, 2006; Pantaleo and Koup, 2004). The hesitation with measuring immune endpoints for large vaccines trials stems both from lack of knowledge that these measurements correlate with protection or

Table 14.2 Assays for use in endpoint assessment for HIV-1 vaccine trials

Test	Advantages	Disadvantages	Ability to distinguish vaccinated from infected	Ability to track disease progression and or vaccine efficacy
Standard EIA	Easy, cheap, rapid, standardized, high-throughput, FDA approved.	Cannot detect early infection reliably	No	Poor
HIV SELECTEST	Easy, cheap, can detect early infection	Not FDA approved	In most cases	Poor
Western Blot	Cheap, fairly standardized	Subjective interpretation of data.	Not always	No
Standard nucleic acid test (NAT) for viral load	Standardized, FDA approved, can be high-throughput, quantitative	Expensive, technologically complex	Yes	Yes
New generation NAT	Standardized, FDA approved, can be high-throughput. Accurate identification of early infection, quantitative	Expensive, technologically complex	Yes	Yes
CD4 T cell count and or percent	Standardized, FDA approved	Expensive, cannot detect early infection reliably	No	Yes
Host genetic factors	Can be high-throughput and provide wealth of information	Exploratory assays, expensive	No	Yes, but unknown for many host genes
Immune function assays	Can provide quantitative multi-parameter information	Time consuming, expensive, technologically complex	Not at present	Yes

clinical benefit, and also lack of confidence to accurately track such immune responses (Garber et al, 2004; Pantaleo and Harari, 2006; Pantaleo and Koup, 2004; Betts et al, in press; Gotch et al, 2005; Klausner et al, 2003).

Recent research efforts have yielded important new correlates of vaccine efficacy or lack of disease progression. Studies in the SIV/macaque model indicate that the major site of early viral replication as well as the bulk of CD4+ T cell depletion occurs on a massive scale, predominantly at intestinal mucosal surfaces, within a short period after the onset of infection. Consequently a prophylactic vaccine must induce immune effectors at mucosal surfaces that act quickly and efficiently to avert the massive depletion of CD4+ T memory cells during primary infection and place constraints on subsequent adaptive immune response in a way that would prevent the rapid evolution of viral variants that escape recognition by CD8+ CTL and antibodies (Betts et al, 2006; Letvin et al, 2006; Mattapallil et al, 2006; Wilson et al, 2006). In addition, systematic efforts have been made to standardize and validate cell-based assays (Cox et al, 2005b; Janetzki et al, 2005a; Mwau et al, 2002; Russell et al, 2003; Samri et al, 2006; Trigona et al, 2003). Table 14.3 lists selected HIV-1 vaccines trials and types of endpoint assays that have been employed.

Table 14.3 Selected trials with immunology endpoints used

Trial[a]	Phase and location	Immunogenicity/ Endpoint assays	Outcome if known
Completed trials			
HVTN trials; 1990–2000	I–II U.S. and S. America	CTL, LPA, neutralization and ELISA	Safe and varying immunogenicity
HVTN, VRC, USMHRP, IAVI and Merck trials; 2000–present	I–II US and S. America	IFNγ ELISpot, IL-2 and IFNγ ICS, neutralization and ELISA	Safe and varying immunogenicity
Vaxgen	III US and Thailand	ELISA and neutralization	No protection
Ongoing or projected trials			
Canarypox + Vaxgen	III Thailand	To be determined	Ongoing
PAVE and Merck	II and IIB US, S. America and Africa	IFNγ ELISpot, IL-2 and IFNγ ICS, central memory T cells	Ongoing

[a] HVTN; HIV-1 Vaccine Trial Network; VRC, Vaccine Research Center, NIH; USMHRP, US Military HIV-1 Research Program; IAVI, International AIDS Vaccine Initiative: PAVE, Partnership for AIDS Vaccine Evaluation (consortium of HVTN, USMHRP and IAVI).

14.5 Methods to assess T cell immunogenicity

Early vaccine candidates focused on products that elicited neutralizing antibodies. However, vaccine-elicited antibodies were unable to neutralize primary virus isolates. At the same time more data were becoming available on the importance of the cellular immune response in controlling HIV-1. Consequently the focus switched to vaccine candidates that elicit either CD4+ and CD8+ T cell responses or CD8+ T cell responses alone.

CD8+ T cells play an important role in controlling HIV-1 and SIV disease progression. Evidence of this is based on several important observations and correlative studies. First, their presence during SIV-infection leads to decreased viral replication and slower disease progression in rhesus macaques (Jin et al, 1999; Schmitz et al, 1999). Second, resolution of acute viremia is coincident with a major expansion of HIV-1-specific CD8+ T cells (Koup et al, 1994; Borrow et al, 1994). Third, during primary and chronic infection, viral variants arise to escape HIV and SIV-specific CD8+ T cells (Goulder et al, 1997; Evans et al, 1999; Allen et al, 2000). Finally, there are strong correlations between expression of certain HLA class I alleles, lack of escape, and nonprogressive HIV-1 infection (Migueles et al, 2000; Kaslow et al, 1996; Scherer et al, 2004). Three attributes of HIV-1-specific CD8+ T cells could be involved in controlling HIV-1 viral replication: magnitude of response, breadth of response and functional quality. The most recent published work suggests that the quality of the CD8+ T cell functional response such as perforin expression, proliferative capacity and cytokine expression, rather than the magnitude or breadth of epitope recognition that correlate with vaccine efficacy (Betts et al, 2006).

In addition to CD8+ T cells, HIV -specific CD4+ T cells play an essential role in maintenance of effective immunity (Picker and Maino, 2000). Recent findings suggest an essential role of CD4+ T helper cells in generating memory CD8+ T cell responses capable of rapid expansion upon secondary antigen challenge (Janssen et al, 2003; Shedlock and Shen, 2003; Sun and Bevan, 2003). The detection of enhanced proliferative responses in individuals who maintain long-term control of HIV-1 replication (Rosenberg et al, 1997), preferential infection of these cells by HIV-1 (Douek et al, 2002) and the maintenance of strong proliferative responses in treated acute infection all provide suggestive evidence that these cells influence viral set point in chronic infection, likely mediating this effect by influencing CD8+ T cell and antibody responses. Vaccine-mediated induction of HIV-1-specific CD4+ T help can reconstitute functional properties of HIV-1 specific CD8+ T cells in vitro (Lichterfeld et al, 2004). In contrast, the ability of CD4+ T cells to proliferate in response to viral antigenic challenge is impaired in persons with progressive infection and high viral loads. Recent studies have indicated that antigen-specific CD4+ T helper cells capable of producing IL-2 may be a key of effective immunity (Harari et al, 2004). Therefore, an effective anti-HIV-1 vaccine should be able to elicit not only effector CD8+ T cell and humoral immunity, but a sizeable CD4+ T cell response.

As with the diagnostic assays described in the section Key components of a laboratory program, the number and type of assays for measuring immune function in vaccine trials has increased progressively throughout the years. For measuring T cell responses, assays such as the traditional semi-quantitative cytotoxic T cell (CTL) and lymphoproliferative assays (LPA) were used. The major limitations of the previous assays such as chromium release CTL assays or LPA assays reside in their qualitative nature, the requirement of elaborate procedures to discriminate effector populations, and, sometimes, the need for fresh PBMCs (Cox et al, 2005a; Gotch et al, 2005). More recently the tetramer, ELISpot and ICS assays have been employed. The advantages of the tetramer, ELISpot and ICS assays include an ability to directly measure the frequency of T cells circulating in the blood, amenability to use with cryopreserved samples, ability to test samples in batches, as well as an opportunity to perform retrospective immunogenicity studies. The ICS and tetramer assays additionally offer the advantage of being able to directly phenotype the responding populations along with measuring their ability to secrete cytokines, respectively. Each assay has its limitations, since the main effector mechanism that correlates with HIV-1 acquisition or progression to disease remains undefined (Garber et al, 2004; Pantaleo and Harari, 2006; Pantaleo and Koup, 2004; Betts et al, 2006; Gotch et al, 2005; Klausner et al, 2003; Roederer et al, 2004). A summary of the old qualitative assays and of the new generation of assays is reported in Table 14.4. This gives us a platform of choices: quantitative assays for assessing vaccine immunogenicity as well more sophisticated, multi-functional assays to identify a correlate of vaccine efficacy.

Two immunologic assays with the greatest utility in evaluating T cell responses elicited by HIV-1 vaccine candidates are the IFN-γ ELISpot and the ICS. These assays are capable of measuring both quantitative and qualitative parameters of antigen-specific T cell responses. If positive data are obtained with either of these assays with specimens from a vaccine trial,

Table 14.4 Assessment of T cell mediated immune responses

	Old generation assays[a]		New generation assays[b]			
	^{3}H-LPA	^{51}Cr-CTL	ELISpot[c]	ICS	CFDA-SE	Tetramer
CD4	Y	N	Y	Y	Y	N
CD8	Y	Y	Y	Y	Y	Y

[a] qualitative results and loss of sensitivity with cryopreserved samples.
[b] quantitative results without loss of sensitivity with cryopreserved samples.
[c] depending on Ag (i.e. 15–18mer peptides) and cytokine (i.e. IL2 vs. IFNγ).

a second phase of analysis is performed that includes determinations of epitope specificity and breadth of T cell responses. This type of information is important to determine if vaccine can induce responses that can recognize virus of multiple subtypes and in various geographic regions. This information will guide decisions about whether a single vaccine will work globally or suggest steps to augment its approach.

The sensitivity of the new generation assays has been investigated by several groups (Sun et al, 2003; Hobeika et al, 2005). One common aspect of these assays is that they require cell samples with viability of greater than 66% for the ELISpot and 70% for the flow-based assays. The required viability is influenced by specimen processing and storage of the samples, as previously discussed, and also by the conditions used during cell thawing (Disis et al, 2006). When these assays are used to support vaccine licensure, they should be validated based on the guidelines provided by the International Conference on Harmonization of Technical Requirements for Registration of Pharmaceuticals for Human Use (ICH): Text on Validation of Analytical Procedures (http://www.ich.org/LOB/media/MEDIA417.pdf) and on the FDA "guidance for Industry: Bioanalytical Methods Validation" (http://www.fda.gov/CDER/GUIDANCE/4252fnl.pdf). These documents define eight parameters in assay validation: specificity/selectivity, accuracy, revision, detection limit, quantification limit, linearity, range and robustness. A detailed description of assay validation is beyond the scope of this chapter and have been addressed elsewhere (Janetzki et al, 2005b; Perfetto et al, 2004; Maecker et al, 2005). In addition, key parameters in the ELISpot and flow-based assays that should also be monitored to comply with Good Laboratory Practices (GLP) requirements are discussed in Section 14.6.

ELISpot

Recently, the IFN-γ ELISpot assay has been validated (Russell et al, 2003; Mwau et al, 2002) for use in clinical trials (see Figure 14.1). This assay uses 96-well plates onto which is coated a monoclonal antibody that can capture human IFN-γ secreted by T cells after incubation with the appropriate peptide antigens. After removing the cells, a biotinylated secondary antibody specific for IFN-γ is added, followed by a detection system to visualize the antibody-IFN-γ-antibody sandwich. Bound cytokine is visible through color development. Areas of color development form spots that represent single cells that have secreted cytokine in response to antigen recognition.. The spot number per well denotes an estimate of the frequency of antigen-specific T cells.

For the ELISpot assay, independent of the detection cytokine, the main parameters to be monitored during the testing of samples for clinical trials are indicated in Table 14.5. The most suitable approach to quality control

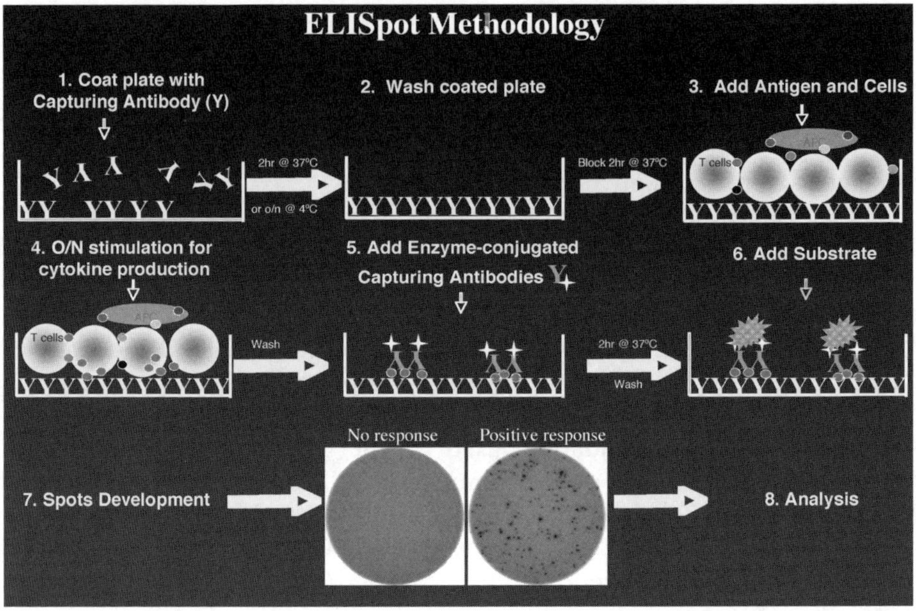

Figure 14.1 Diagram of ELISpot methodology. In the enzyme-linked immunosorbent spot assay, cytokine-specific antibodies absorbed onto the membrane of a 96-well plate capture the cytokine secreted by antigen-stimulated cells. After cell removal, enzyme-linked cytokine-specific antibodies recognize the cytokine and complete the "sandwich" can be visualized as a "spot" by the addition of the appropriate substrate. Each spot represent an individual antigen-specific cell and results are reported as Spot Forming Cell (SFC) per 10^6 PBMC

the assay is to use a PBMC sample collected from a normal donor with known reactivity against a positive control such as the CEF pool which includes a variety of optimal HLA-restricting peptides representing epitopes in CMV, EBV and Flu (Currier et al, 2002), or the peptide pool representing the hCMV pp65 protein which can detect both CD4 and CD8 responses (Maecker et al, 2001) (see Table 14.5). The use of such positive PBMC controls also allows standardization of the reader during the QC of the results. Moreover, the technical operators could use the analysis of the longitudinal positive control data to establish acceptance criteria for inter-assay variation within the laboratory. The use of such a positive sample can also provide information on the impact of the thawing process on the cells by recording cell recovery and viability data. Moreover, such positive controls with defined pattern of antigen recognition of cell samples and reagent are instrumental in performing bridging studies to a new lot of reagent. It is also useful to include control wells without cells, but only

Table 14.5 Key factors in ELISpot assay

Factor	Explanation
Medium composition	Reduce non specific activation of cells
Cell/well	Optimize contact between cells and capture antibody
Pre-wetting of membrane	Optimize coating of membrane
Antibody concentration	Optimize spot size
Substrate concentration	Avoid increased background of membrane
Reading parameters	Optimize reading conditions and exclude of artifacts

reagents to control for possible artifacts that are produced by precipitation of the reagents due to the cell medium or the reagents themselves.

Polychromatic flow cytometry

Recent technological advances have enabled simultaneous measurements of multiple T cell functions. Of the two flow-based assays that are considered in this section, the ICS assay can quantitatively assess the frequency of antigen-specific T cell subsets (i.e. CD4+ or CD8+) following *in vitro* stimulation (Perfetto et al, 2004) (see Figure 14.2). Briefly, cryopreserved T cells are thawed and rested overnight before stimulation with appropriate peptide pools. The stimulation is conducted in the presence of agents that can block the release of cytokine or other molecules from the Endoplasmic Reticulum or Golgi apparatus during incubation. The samples are subsequently permeabilized and stained with fluorophore-conjugated antibodies directed against phenotypic cellular markers or intracellular cytokines.

A second flow-based assay utilizes carboxyfluorescein diacetate, succinimidyl ester (5, 6- CFDA-SE) a non polar molecule that penetrates cell membranes and is converted into anionic CFSE (Lyons and Parish, 1994). CFSE binds irreversibly to available amine groups of cellular proteins. When cells divide, CFSE labeled proteins are equally distributed between the daughter cells. Therefore, a halving of cellular fluorescence intensity is present in each successive cell generation in a population of proliferating cells. The proliferative capacity of antigen responding cells is detected as the number of cell generations that occur during a 6–7 days *in vitro* stimulation with antigens. At this time the cells can be also stained with antibody that recognizes phenotypic surface T cell antigens such as CD3, CD4, and CD8.

Flow cytometric-based assays are moving towards validation for multi-functions and phenotypic parameters of antigen-specific T cell subsets. The main difficulty in validating this assay resides in the validation and daily quality control of the instrument that must precede the analysis of the results. We refer to the following references (Perfetto et al, 2004; Shapiro,

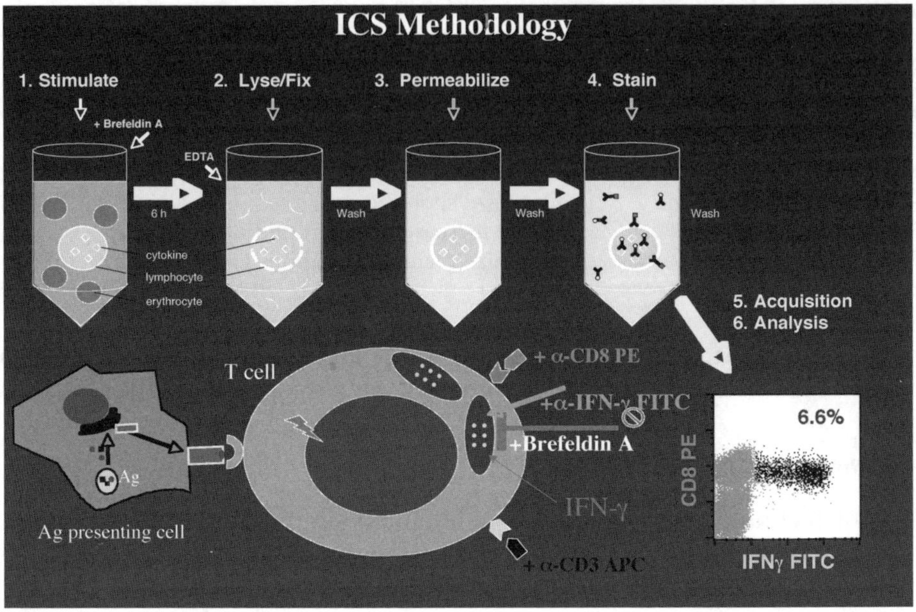

Figure 14.2 Diagram of ICS methodology. Antigen-specific stimulation of PBMC, in the presence of agents that can block the release of molecules from the Endoplasmic reticulum or Golgi apparatus, leads to the intracellular accumulation of cytokines. Membrane permeabilization of the cells allows the entry of anti-cytokine specific fluorophore-conjugated antibodies that can stain the cells responding to the antigen stimulation and allows their enumeration using a flow cytometer. Simultaneous staining for phenotypic markers allows the identification of the antigen-specific functional memory and effector lymphocyte subsets

2003) for guidelines on how to reduce changes in the instrument settings that could influence the signal to noise ratio and voltages required to identify the different fluorophores included in the staining panel. The consistent performance of the instrument will facilitate the data acquisition and analysis in a reliable and reproducible manner (Maecker et al, 2005).

Another difficulty in ICS performance is the choice of staining reagents for cellular surface antigens and intracellular cytokines to define phenotype and function of the cell subset of interest. The choice of reagents can be difficult because of the relative expression of the antigen/cytokine and intensity of the fluorophore to detect them, the complexity in developing custom conjugate antibodies, the labor intensity and expense required to titrate the conjugated antibodies, and lastly, the necessity to configure the laser and filter of the instrument. Once the details of the stimulation, staining, data acquisition and gating procedures are fully standardized and validated, the key parameters that must be monitored are indicated in Table 14.6.

Table 14.6 Key factors in flow-based assay

Factor	Explanation
Medium composition	Reduce non specific activation of cells
Number of Cells/test	Optimize number of events required to define positive responses
Antibody concentration	Optimize staining and subset separation
Daily Instrument QC and set-up	Verify laser performance and fluidic operation

Medium composition, number of cells per test, and concentration of antibody can always be evaluated if a cell sample with known reactivity against one antigen is tested in each assay along with the clinical samples. Longitudinal data with the unstimulated control and antigen-stimulated condition can be informative of background and antibody staining.

Selection of peptide antigens

There are two major considerations when performing T cell assays: the first is to accurately define the immune response to a candidate vaccine by use of peptides that match the vaccine candidate. The second is to define the breadth of immune responses that could be potentially relevant to protection within a particular population. This goal is best accomplished by selecting peptide reagents to depict the epitopes that T cells may encounter following HIV-1 exposure and to use these peptides to evaluate a vaccine candidate's potential to induce immune responses against circulating viruses. Considering the enormous diversity of certain HIV-1 gene products, a standardized peptide panel is needed that satisfactorily represents circulating HIV-1 strains to which these individuals are exposed. However, with the increasing recognition of superinfection and genetic recombinants, limiting evaluation to peptides from a single reference strain, particularly in sites with increasing complexity of circulating HIV-1 strain mixtures may be insufficient. Vaccine developers are also facing this issue in assessing immunogenicity to candidate vaccines containing gene inserts from various subtypes and genes.

Currently three types of peptide antigens strategies are being evaluated: use of consensus sequence artificially derived peptides (Gaschen et al, 2002); peptides based on centrally-derived computational algorithm to define potential T cell epitope (PTE) sequences; (Li et al, in press a); and toggled peptides which represent the conserved peptide as well as additional toggled peptides where certain amino acids in variable positions of a protein sequence are substituted and occur frequently in the HIV sequence database {Brander, 2006 145 /id}. Regardless of the strategy used to determine the sequence of the peptides, the peptides are usually synthesized as 15 amino acid length and overlap by 11 residues. Because of the large number of peptides used to detect vaccine-induced responses, they are usually tested as pools ranging from 50 to 125 peptides per pool. Peptides synthesis and pooling process are usually performed by the manufacturer because of the ease and simplicity associated with pooling lyophilized peptides rather than individual solubilized peptides.

Further, to avoid contamination or errors during pooling, an additional quality control check is recommended on the peptide pool for concentration and purity prior to use in the cellular assays. This testing should be performed by an entity independent from the manufacturer. The size of the peptides and the number of peptides per pool make it impossible to verify the presence of every peptide after pooling. Rather, it is best to randomly test at least 10% of the pools and to confirm peptide pool composition by HPLC. As contamination of the peptide pools could cause specific activation of the PBMC, it is recommended to use a panel of PBMCs collected from individuals with known positive and negative reactivity to a selection of the antigens represented by the peptide pools to evaluate the performance of the different lots of reagents and the incidence of false positive responses.

14.6 Humoral immune responses to HIV-1

Humoral immune responses are mediated by antibody molecules, which are matured and functional immunoglobulin (Ig) proteins produced by specialized, differentiated B lymphocytes called plasma cells. Antibody molecules have two separate functions: to bind specifically (via the antibody's variable or V region) to antigens from the pathogen that elicited the immune response; and to recruit cells and molecules to destroy the pathogen once the antibody is bound to it (Janeway and Travers, 1997). In response to HIV-1 infection, antibodies may be directed at several different HIV-1 structural or regulatory proteins. While a declining antibody titer to the HIV-1 gag protein has been shown to be associated with HIV-1 disease progression (Schmidt et al, 1987; de et al, 1989), the primary focus for study of HIV-1 humoral responses has been placed on antibodies to the envelope (env) proteins (the gp160 precursor and the gp120 surface and gp41 trans-membrane proteins). These antibodies are presumed to be the major effectors of a functional response, as they bind to the surface of the virion assumed to be the primary target by which to neutralize HIV-1.

Methods to assess HIV-1 binding antibodies

The simplest test to look for antibodies against any pathogen is to assess for the presence of binding antibodies that attach to any of the antigens of the pathogen. The reader is referred to the Section 14.3 titled Clinical diagnostic and endpoint tests for a detailed description of the EIA and Western blot assays in the context of HIV-1 clinical trials.

Methods to assess HIV-1 neutralizing antibodies

When different populations of antibodies are produced in response to a pathogen or virus, there are a number of different possible functional outcomes. Antibodies that modify or inhibit the pathogen or its clinical course, are functional and proposed functions for pathogen-associated antibodies are listed in Table 14.7.

Various assays are used to measure antibodies that neutralize HIV-1 (Pantophlet and Burton, 2006). These assays rely on different technologies, but are based on defining a reduction of viral target cell infection by antibodies (polyclonal or monoclonal) in cells that express suitable fusion receptors for virus entry. The mechanisms of HIV-1 neutralization are poorly understood. While antibodies are known to play an important role in protection in viral diseases such as polio (Hammon et al, 1953), rabies (Suss and Sinnecker, 1991), measles (Chen et al, 1990), and influenza (Frank et al, 1983), the role of neutralizing antibodies (NAbs) in HIV-1 protection and pathogenesis requires further definition. However, passive administration of one or more broadly neutralizing HIV-1 monoclonal antibodies, prior to challenge with SIV–HIV hybrid virus prevents infection (Mascola et al, 2000; Nishimura et al, 2002; Nishimura et al, 2003; Parren et al, 2001). In contrast, administration of these antibodies to chronically infected humans that underwent anti-retroviral interruption, showed delayed viral rebound, thought the virus did escape from the antibody-induced immune pressure (Trkola et al, 2005).

The generation of an antibody response capable of neutralizing primary isolates of multiple HIV-1 subtypes is a desired characteristic for candidate HIV-1 vaccines. The design of novel envelope subunits that are presented to the immune system in physiologically relevant structures that more closely mimic the trimeric conformation of the virion envelope is ongoing (Li et al,

Table 14.7 Antibody functions in pathogen infections

1. Lysis of pathogen/virus
2. Lysis of cells infected by the pathogen (example: antibody dependent cellular cytotoxicity, ADCC)
3. Diversion of antibody-coated pathogen (opsonized) to antigen presenting cells via cellular Fc (constant fragment) receptor attachment to the Fc portion of the antibody molecule
4. Clearance of cellular debris post-infection
5. Inhibition or blocking of pathogen infectivity, commonly called "neutralization"

2006). These approaches have included the expression of native oligomeric and monomeric envelopes (Li et al, 2006), expression of multiple envelopes using DNA vectors followed by protein boosting (Pal et al, 2006; Wang et al, 2006), and production of envelopes mutated to induce stable, cleaved, trimeric conformations (Beddows et al, 2006). An increasing number of Phase I and II human trials that are being conducted with these new envelope designs, either alone, or in combination with T cell-based vaccines.

Several laboratories have worked towards developing high throughput, sensitive, and reproducible neutralization assays that are validated for use with human trial samples. Numerous *in vitro* neutralization assay formats have been developed, with multiple variables and endpoints. Table 14.8 indicates the most commonly utilized assay formats and a brief overview of a neutralization assay schematic (Mascola et al, 2005).

In most neutralization assays, HIV-1 and antibody are incubated together and then added to target cells that express CD4 and one or more of the major HIV-1 coreceptors (CXCR4 and CCR5). Infection of the target cells results in a quantifiable readout in the cultures, for example, HIV-1 p24 gag protein production, HIV-1 RT enzymatic activity, syncytium formation, or activation of a reporter gene such as luciferase or green fluorescent protein (GFP). The early, traditional neutralization assays employed transformed T lymphocyte cell lines or primary peripheral blood mononuclear cells (PBMC) derived from the blood of HIV-1-seronegative donors (Mascola, 1999). The use of membrane permeabilization to allow fluorescent-labeled antibodies to enter cells and bind to proteins expressed intracellularly, followed by flow cytometric detection of cells expressing a marker viral protein such as p24 gag allows enumeration of infected PBMC as a highly specific endpoint for HIV-1 neutralization using primary cells (Darden et al, 2000; Mascola et al, 2002).

Recently, a neutralization assay format that exploits molecular clones of the HIV-1 genome, HIV-1 env clones, and cells engineered to over-express CD4 and HIV-1 coreceptors, as well as an HIV-1-inducible reporter gene, has been advocated as an initial platform to standardize measurement of HIV-1 NAb for vaccine trials (Mascola et al, 2005). In this assay (described briefly in Table 14.8), an engineered virus (called a pseudovirus (PSV)) is first created. The plasmid (DNA) expressing env is introduced into the cell together with a plasmid that expresses all of HIV-1 minus the env. The PSV stock will be made in the producer cells (293T cell line) such that the viruses will express the env at the exterior and the viral genome packaged in the PSV will have a defective env, so no other infectious particles will be produced. This way, the PSV can be used to infect cells expressing a reporter gene, but no new infectious viruses will be made. The reporter gene is expressed only when HIV-1 infects the cell. Upon entry, the HIV-1 tat protein signals the expression of a reporter (i.e. the luciferase

Table 14.8 Neutralization assay formats

Viruses	Target cells	Endpoint readouts
T Cell Line Adapted (TCLA)	CD4+ T Cell Lines	Syncytia, plaques p24 antigen, RT enzyme, Cell killing
Primary Isolates	PBMC	p24 antigen, intracellular p24 RT enzyme
Primary Isolates Pseudoviruses TCLA	Cell lines with CD4, coreceptor(s), and an inducible reporter gene	Luciferase activity, green fluorescent protein, beta-galactosidase activity, others
Neutralization Schematic:		
Virus + AB – – – – – – > (30 min–18 hours)	Add Target Cells – – – – > (2–7 Days)	Measure Infection

(adapted from Mascola et al, 2005)

enzyme), and the enzymatic activity will then be present in the target cells. If substrates for luciferase are added to the cells, the luciferase enzymatic activity will catalyze the production of light, which can then be quantified using a luminometer. Because no more new particles of HIV-1 can be made by the target cells (the genome has a defective env), this is called a single cycle of infection assay.

In a neutralization assay, the PSV stock can be mixed together with various samples of serum or plasma from HIV-1+ patients or HIV-1 vaccine recipients, and antibodies directed against the env may bind to the virus. The virus/antibody mixtures are then added to the target cells; if there are NAb in the sera or plasma samples, then the virus will be blocked from infecting the cells. When substrates are added, there will be little or no light emitted; this is a quantifiable readout for neutralization (over nearly a 6 log range). The light emitted when NAb are present is compared to the light emitted when no NAb are mixed with the PSV, and a percent inhibition of infection, or percent neutralization is calculated. The comparative properties of this type of reporter cell assay utilizing PSVs, and the PBMC based assay format are listed in Table 14.9.

Recently, the GHAVE (Gates HIV/AIDS Vaccine Enterprise) Humoral Immune Assay Standardization Committee came to a consensus that the PSV assay will be a good platform to standardize neutralization methodologies amongst multiple laboratories. This standardization will facilitate comparisons of vaccine products and will provide a starting place for the GLP-compliant assessment of vaccine-induced NAb (Mascola et al, 2005). A proficiency test that included 17 laboratories from five different countries, to assess the reproducibility of PSV neutralization assays across multiple sites has been conducted. Preliminary data show good consistency between laboratories, however, greater stringency in assay conditions will be needed to obtain more concordant data (Li et al, in press b). Due the fact that

Table 14.9 Characteristics of primary isolate/PBMC assays versus Pseudovirus assays

Characteristic	PBMC assay	Pseudovirus assay
Common Readout	Extracellular p24	Reporter gene activity
Physiologic target cells?	Yes	No
Target Cell variability	High	Low
Sensitivity	Low to Moderate	High
Reproducibility	Poor	Excellent
Quantitative range	Moderate 2–4 logs	High (up to 6 logs)
Throughput	Moderate	High
HIV-1 Envelope source	Mixed quasispecies	Single or multiple clones
Assay validated?	No	Yes

cloned viruses and neoplastic cell lines, which may be non-physiologic, are used in the PSV assay, many investigators in the research arena believe that both primary cell (lymphocytes, dendritic cells, and macrophages) (Holl et al, 2006), as well as cell line-based assays, should continue to be evaluated in parallel. This will allow for generation of comparative data, regardless of assay choice, and enable informed decisions about the potential of immunogens to elicit quality NAbs.

Additionally, standardized panels of Env-pseudotyped viruses to assess the potencies and breadths of Nabs elicited by vaccine immunogens have been established. A three tier system to assessing the neutralizing antibody responses generated by vaccine candidates has been established (Li et al, 2005; Li et al, in press b). Tier 1 represents virus strains present in the vaccine and some neutralization-sensitive viruses not included in the vaccine. Tier 2 is a panel of heterologous viruses matching the genetic subtype(s) of the vaccine. Tier 3 is a multi-clade panel comprised of viruses of each genetic subtype, excluding the genetic subtype evaluated in Tier 2. The data generated using the standardized virus panels and validated assays will allow comparisons of the breath and potencies of new envelope-based immunogens and enable quantitation of elicited NAbs.

14.7 Operating in a Good Clinical Laboratory Practices (GLP) environment

GLP elements of a laboratory program

GLP, or Good Clinical Laboratory Practice, is a set of guidelines that strives to promote the quality and validity of the data. The goal of GLP is to ensure that the results of a study involving human subjects are reliable, repeatable, auditable and recognizable by countries worldwide. GLP derives from the merging of two sets of regulations: Good Clinical Practice (GCP) and Good Laboratory Practices (GLP). GCP is an international, ethical, scientific quality standard for designing, conducting, recording and reporting trials that involve human subjects. Compliance with this standard provides public assurance that the rights, safety and well being of trial subjects are protected (ICH, 1996; FDA, 2001; Code of Federal Regulations, Title 42; Department of Health and Human Services, 2004; BARQA, 2003). GLP is also a managerial concept covering the organizational process and the conditions under which laboratory studies are planned, performed, monitored, recorded and reported (Code of Federal Regulations, Title 21). Presently no standards or regulations exist for conducting immunogenicity assays for clinical studies. GLP compliance is a continuous process where the laboratory staff strives to obtain accurate data with minimum incidents of false negative and false positive data. Moreover, because Investigational New Drug or Biologics Sponsors of such clinical trials utilize immunogenicity data from early clinical trials to guide product advancement decisions, the Food and Drug Administration guidelines for GLP and appropriate measures for quality control and assurance for non clinical studies should be adhered to when performing immunological endpoint assays in support of vaccine licensure.

GLP training of personnel

One of the first steps to create a GLP compliant environment in a laboratory setting is to provide GLP training to all personnel involved in the studies, from Study Directors to Laboratory Operators. This step is particularly critical when GLP regulations are to be introduced in an academic environment. Each laboratory conducting endpoint assays in a GLP compliant environment implements training standard operating procedures (SOPs) to evaluate the training and competency of personnel performing the studies. Personnel must satisfactorily complete the institutional general laboratory safety training, pathogen specific training and the facility safety training before being trained to perform laboratory assays. These safety trainings are to ensure operator safety and also to ensure that all material is handled and disposed of in the proper way, and that equipment is maintained in a safe and clean manner. All laboratories performing GLP compliant studies must conduct routine competency and proficiency testing for their employees to ensure that the operator possesses the ability to perform the assay as stated in the SOP. Furthermore, external proficiency panels obtained through an external accreditation body (i.e. CLIA, CAP, etc.) can be used to compare data generated at multiple sites.

Standard operating procedures

SOPs are another critical element of a GLP compliant study. SOPs describe experimental and organizational procedures, equipment and facilities use, reagents and specimen handling, training, data management, and archival procedures. The use of SOPs ensures that procedures are performed the same way, every time they are applied. SOPs must be periodically reviewed, kept current, and their distribution must be controlled to ensure use of the latest version. Any procedure that differs from an SOP must be recorded, reported and filed as a deviation.

Reagents and equipment

Reagents necessary for a study protocol must be produced or purchased in quantities sufficient for the analysis of test samples. It is important to document the quality of the reagents and to perform bridging experiments before a reagent lot expires or is exhausted, since some reagents are subject to lot-to-lot variation. Optimal storage conditions should be maintained to ensure the integrity of the reagents for the duration of the study protocol. Upon receipt of a reagent, the Certificate-of-Analysis (COA) and a Material Safety Data Sheet (MSDS) must be filed along with any other information regarding the purity and composition of the reagent. The equipment used in a GLP compliant study must be properly designed, installed and qualified for operation and performance, and it must be calibrated on a routine basis. Records of such maintenance are to be retained. Each item of equipment used in GLP compliant studies must be inventoried and labeled to differentiate it from non-GLP compliant ones.

Reporting requirements

Each endpoint assay protocol study requires an approved Study Plan in support of GLP compliance. It is the responsibility of the Study Director to write a Study Plan, which must be implemented before study initiation. The Study Plan provides evidence that a study was planned thoughtfully. It should describe how specimens are acquired and tested, provide a timeline of testing and define how the data is to be statistically analyzed and reported. A Study Plan must be current and it must conform to an approved Study Protocol and its appendices: a Study Plan is an absolute requirement in both the FDA and Organization for Economic Co-operation and Development (OECD) regulations for GLP.

At the end of a study the Study Director needs to write a Final Study Report. The Final Study Report identifies the facility performing the study and the date on which the study was initiated and completed, the name of the Study Director, the objectives and procedures stated in the approved

Study Plan, including any amendments, the statistical analysis of the data, and the conclusions obtained from the analysis, a description of the methods used, a description of all deviations and whether they affected the integrity of the study, the locations where all specimens, raw data, and the final report are to be archived, a QA Statement indicating the phases of the study inspected by the QAU and when the results of the inspections were submitted to the Study Director and Management. The final report must be signed and dated by the Study Director.

Internal and external audits

The internal QAU is an entity independent of the conduct of the study, it is responsible for ensuring that the study performed is compliant to GLP regulations. The assessment process (audit) conducted by the internal QAU is crucial for maintaining the integrity of a study. The purpose of an audit is to ensure that procedures are being followed accurately, data is being recorded indelibly, and organization is being maintained throughout the laboratory. Internal audits should be conducted by the QAU at intervals following a predetermined program. Audits cover, but are not limited to the following items: maintenance of SOPs, adherence to SOPs, respect of safety procedures, inspections of QC logs, maintenance of reagent logs, maintenance of raw data, maintenance of equipment repair logs, maintenance of personnel/training files, and any other aspect of the laboratory that may be subject to audit by an external source or government agency. Following the conclusion of an internal audit, the QAU submits a report of non-conformances to the Study Director for comments and responses. The Study Director must respond to the issues addressed in the audit report within a predetermined length of time. The QAU checks for corrective actions to listed non-conformances and their proper documentation. The QAU compiles trend line results for quality controls and corrective actions into a report that is forwarded to Management.

The external audits are necessary to ensure that all aspects of the studies are GLP compliant, including the use of validated assays and the functions of an internal QAU. In addition, the external auditors, because of broader experience from auditing multiple laboratories in various institutions provide assurance that the laboratory standards of operation are conformant to those of other national and international laboratories.

Maintaining a GLP environment

To maintain a GLP environment for a study protocol it is critical that all of the elements pertinent to GLP regulations are in place and operational. These elements include organization and personnel, facilities, equipment, specimen and reagents, Study Plans, SOPs, QAU, data management unit, and a system of storage and retention of records. The most appropriate way to ensure that the GLP regulations are applied is to routinely audit all aspects of a GLP environment. It is the responsibility of an internal QAU to audit and document all issues of non-compliance, and to ensure that corrective actions are taken in a timely fashion. Audits of the trial facility by an external QAU should also occur to monitor all aspects of the GLP compliant facility, including the internal QAU.

External quality assurance (EQA) programs

There are multiple organizations conducting HIV-1 vaccine trials globally (McMichael and Hanke, 2003). In the absence of a single central laboratory that can perform endpoint assays, it is imperative that the data from multiple laboratories performing assays in support of a single vaccine candidate or even multiple vaccine candidates, are reliable and reproducible so that immunogenicity data can be effectively compared and deliberate decisions occur about which candidates to advance into efficacy trials (Klausner et al, 2003). This will require standardized assays, common reagents and SOPs, highly trained technologists conforming to the GLP guidelines and the implementation of internal and external QA programs as a tool by which to measure and monitor laboratory performance.

An EQA program for immune monitoring assays is necessary to ensure competency among multiple laboratories and it serves three purposes: (1) it provides an internal measurement tool for ensuring that the information a laboratory generates and provides is accurate, timely, clinically appropriate and useful; (2) it provides the sponsoring and regulatory agencies with confidence that individual laboratories are generating data with a rigor that will support vaccine licensure; (3) it ensures the clinical trial volunteer that the system is working together to provide accurate and reliable information. Most of the HIV-1 diagnostic assays described in this chapter are covered by EQA programs administered through the College of American Pathologists, the Clinical and Laboratory Standards Institute and other organizations. Until recently no EQA programs existed for cellular or humoral immunogenicity endpoint assays. For the first time the advent of standardized quantitative, single cell assays to measure T cell responses such as the tetramer, ELISpot and intracellular cytokine staining assays can be used to develop proficiency panels of reagents (Pantaleo and Harari, 2006; Gotch et al, 2005).

Through efforts pioneered by the Division of AIDS, EQA programs for ELISpot and ICS have been established for laboratories involved in testing HIV-1 vaccines (Cox et al, 2005b; Maecker et al, 2005). The development of an EQA program to measure ELISpot and ICS assays have included the following steps: (1) establish cell and reagent banks (PBMC positive and negative controls, high/low responders) to define positive and negative controls; (2) establish frequency of proficiency testing plan and statistical methods to evaluate pass/fail criteria for ICS and ELISpot; (3) establish/test reagent shipping and data transfer logistics; (4) blind, code and ship reagents, specimens and protocols to laboratories for proficiency testing; (5) develop an on-line web-based system to post data and; (6) analyze results and notify laboratories of pass/fail status in EQA program. The results of ELISpot and ICS EQA programs have been published, are continually being refined and are becoming open to more participants via commercialization (Cox et al, 2005b; Maecker et al, 2005). The cancer vaccine consortium (CVC, Sabin Vaccine Institute, 2007) has also set up an EQA for ELISpot. Eventually these EQA program will mature and participating laboratories will be assessed bi-annually and obtain feedback on their performance. Ultimately, the application of these standardized and validated immunogenicity assessments utilized under GLP guidelines will provide a platform to directly and quantitatively compare vaccine regimens and doses among different clinical trials, regardless of geographic location.

14.8 Conclusions

Development of a safe and effective HIV-1 vaccine is one element of a multi-pronged program to address the AIDS pandemic. Design of an HIV-1 vaccine remains one of the most challenging and vital areas of biomedical research. New vaccine concepts in efficacy testing are able to generate a T cell immune response that prolongs a symptom-free disease period, but do not prevent infection. Success with these current vaccine strategies will depend upon the generation of large CD4+ and CD8+ T cell responses that span multiple functions, exhibit broad specificity, HIV-1 cross-clade reactivity, and remain durable.

To support the testing of HIV-1 vaccine candidates in human clinical trials requires a state-of-the-art laboratory program that is capable of prioritizing vaccine candidates rationally, as well as building on findings for further development. Beyond the safety analysis, all decisions regarding the advancement of products are based on assays conducted on specimens collected during the trials. Assays must be high throughput, correlative and comply with GLP regulations to ensure reproducible, robust data that gain sponsor confidence as well meet regulatory agency scrutiny.

References

ICH (1994). International conference on harmonization of technical requirements for registration of pharmaceuticals for human use. Text on validation of analytical procedures http://www.ich.org/LOB/media/MEDIA417.pdf.

Code of Federal Regulations, Title 42, Volume 3, Part 493 – Laboratory Requirements. U.S. Government Printing Office via GPO Access, 01-Oct-2001. http://www.access.gpo.gov/nara/cfr/waisidx_04/42cfr493_04.html.

ICH (1996). International conference on harmonization of technical requirements for registration of pharmaceuticals for human use. Validation of analytical procedures: Methodology http://www.ich.org/LOB/media/MEDIA417.pdf.

BARQA (2003). http://www.barqa.com/cms.php?pageid=390&index=1).

U.S. Food and Drug Administration (2001). Guidance for Industry: Bioanalytical Methods Validation. http://www.fda.gov/CDER/GUIDANCE/4252fnl.pdf.

Code of Federal Regulations, Title 21, Volume 1, Part 58 – Good Laboratory Practice for Nonclinical Laboratory Studies Reference. U.S. Government Printing Office via GPO Access, 01-Apr-2005. http://www.access.gpo.gov/nara/cfr/waisidx_05/21cfr58_05.html.

Cancer Vaccine Consortium, Sabin Vaccine Institute http://www.sabin.org.

Addo MM, Yu XG, Rathod A, Cohen D, Eldridge RL, Strick D, Johnston MN, Corcoran C, Wurcel AG, Fitzpatrick CA, Feeney ME, Rodriguez WR, Basgoz N, Draenert R, Stone DR, Brander C, Goulder PJ, Rosenberg ES, Altfeld M and Walker BD (2003). Comprehensive epitope analysis of human immunodeficiency virus type 1 (HIV-1)-specific T-cell responses directed against the entire expressed HIV-1 genome demonstrate broadly directed responses, but no correlation to viral load. *J Virol*, **77**, 2081–2092.

Aldovini A and Walker BD (eds) (1990). *Techniques in HIV Research*. New York, Stockton Press.

Allen M, Israel H, Rybczyk K, Pugliese MA, Loughran K, Wagner L and Erb S (2001). Trial-related discrimination in HIV vaccine clinical trials. *AIDS Res Hum Retroviruses*, **17**, 667–674.

Allen TM, O'Connor DH, Jing P, Dzuris JL, Mothe BR, Vogel TU, Dunphy E, Liebl ME, Emerson C, Wilson N, Kunstman KJ, Wang X, Allison DB, Hughes AL, Desrosiers RC, Altman JD, Wolinsky SM, Sette A and Watkins DI (2000). Tat-specific cytotoxic T lymphocytes select for SIV escape variants during resolution of primary viraemia. *Nature*, **407**, 386–390.

Beddows S, Kirschner M, Campbell-Gardener L, Franti M, Dey AK, Iyer SP, Maddon PJ, Paluch M, Master A, Overbaugh J, VanCott T, Olson WC and Moore JP (2006). Construction and characterization of soluble, cleaved, and stabilized trimeric Env proteins based on HIV type 1 Env subtype A. *AIDS Res Hum Retroviruses*, **22**, 569–579.

Belshe RB, Clements ML, Keefer MC, Graham BS, Corey L, Sposto R, Wescott S and Lawrence D (1994). Interpreting HIV serodiagnostic test results in the 1990s: social risks of HIV vaccine studies in uninfected volunteers. NIAID AIDS Vaccine Clinical Trials Group. *Ann Int Med*, **121**, 584–589.

Betts MR, Nason MC, West SM, De Rosa SC, Migueles SA, Abraham J, Lederman MM, Benito JM, Goepfert PA, Connors M, Roederer M and Koup RA (2006). HIV nonprogressors preferentially maintain highly functional HIV-specific CD8+ T cells. *Blood*, **107**, 4781–4789.

Betts MR, Gray C, Cox JH and Ferrari G (in press). Identification of antigen-specific T-cell-mediated immunity after HIV-1 infection: implications for vaccine control of HIV. *Exp Rev Vaccines*.

Blankson JN, Persaud D and Siliciano RF (2002). The challenge of viral reservoirs in HIV-1 infection. *Annu Rev Med*, **53**, 557–593.

Borrow P, Lewicki H, Hahn BH, Shaw GM and Oldstone MB (1994). Virus-specific CD8+ cytotoxic T-lymphocyte activity associated with control of viremia in primary human immunodeficiency virus type 1 infection. *J Virol*, **68**, 6103–6110.

Brander C, Frahm N, Joijc N, Yusim K, Nickle DC, Linde CH, Hewitt HS, Faircloth KL, Muldoon M, Kadie C, Walker BD, Heckerman D, Mullins JI and Korber B (2006). Capturing HIV viral sequence diversity in *in vitro* immune analyses. Abstract presented at Fifth International Conference, AIDS Vaccine 2006, Amsterdam, The Netherlands.

Buchbinder S and Vittinghoff E (1999). HIV-infected long-term nonprogressors: epidemiology, mechanisms of delayed progression, and clinical and research implications. *Microbes Infect*, **1**, 1113–1120.

Bull M, Lee D, Hallstrom J, Stucky J, Beckham C, Gordan J, Penn J, Baydo R, Deers M, Chiu Y, Rubin A, Horton H and McElrath MJ (unpublished). Defining blood processing parameters for optimal detection of cryopreserved antigen-specific responses for HIV vaccine trials.

Chen RT, Markowitz LE, Albrecht P, Stewart JA, Mofenson LM, Preblud SR and Orenstein WA (1990). Measles antibody: reevaluation of protective titers. *J Infect Dis*, **162**, 1036–1042.

Cox JH, deSouza M, Ratto-Kim S, Ferrari G, Weinhold K and Brix DL (2005a). Cellular immune assays for evaluation of vaccine efficacy in developing countries. In: NR Rose, RG Hamilton and B Detrick (eds), *Manual of Clinical Immunology Laboratory*. Washington, DC, ASM Press, p 301.

Cox JH, Ferrari G, Kalams SA, Lopaczynski W, Oden N and D'souza MP (2005b). Results of an ELISPOT proficiency panel conducted in 11 laboratories participating in international human immunodeficiency virus type 1 vaccine trials. *AIDS Res Hum Retroviruses*, **21**, 68–81.

Currier JR, Kuta EG, Turk E, Earhart LB, Loomis-Price L, Janetzki S, Ferrari G, Birx DL and Cox JH (2002). A panel of MHC class I restricted viral peptides for use as a quality control for vaccine trial ELISPOT assays. *J Immunol Methods*, **260**, 157–172.

Darden JM, Polonis VR, deSouza MS, Chantakulkij S, Brown AE, Birx DL and Pattanapanyasat K (2000). A flow cytometric method for measuring neutralization of HIV-1 subtype B and E primary isolates. *Cytometry*, **40**, 141–150.

de WF, Lange JM, Houweling JT, Mulder JW, Beemster J, Schellekens PT, Coutinho RA, van der, NJ and Goudsmit J (1989). Appearance of predictors of disease progression in relation to the development of AIDS. *AIDS*, **3**, 563–569.

Disis ML, dela RC, Goodell V, Kuan LY, Chang JC, Kuus-Reichel K, Clay TM, Kim LH, Bhatia S, Ghanekar SA, Maino VC and Maecker HT (2006). Maximizing the retention of antigen specific lymphocyte function after cryopreservation. *J Immunol Methods*, **308**, 13–18.

Douek DC, Brenchley JM, Betts MR, Ambrozak DR, Hill BJ, Okamoto Y, Casazza JP, Kuruppu J, Kunstman K, Wolinsky S, Grossman Z, Dybul M, Oxenius A, Price DA, Connors M and Koup RA (2002). HIV preferentially infects HIV-specific CD4+ T cells. *Nature*, **417**, 95–98.

Evans DT, O'Connor DH, Jing P, Dzuris JL, Sidney J, da SJ, Allen TM, Horton H, Venham JE, Rudersdorf RA, Vogel T, Pauza CD, Bontrop RE, DeMars R, Sette A, Hughes AL and Watkins DI (1999). Virus-specific cytotoxic T-lymphocyte responses select for amino-acid variation in simian immunodeficiency virus. *Env Nef Nat Med*, **5**, 1270–1276.

Frank AL, Taber LH, Glezen WP, Geyer EA, McIlwain S and Paredes A (1983). Influenza B virus infections in the community and the family. The epidemics of 1976–1977 and 1979–1980 in Houston, Texas. *Am J Epidemiol*, **118**, 313–325.

Garber DA, Silvestri G and Feinberg MB (2004). Prospects for an AIDS vaccine: three big questions, no easy answers. *Lancet Infect Dis*, **4**, 397–413.

Gaschen B, Taylor J, Yusim K, Foley B, Gao F, Lang D, Novitsky V, Haynes B, Hahn BH, Bhattacharya T and Korber B (2002). Diversity considerations in HIV-1 vaccine selection. *Science*, **296**, 2354–2360.

Gotch F, Holmes H and Imami N (2005). The importance of standardisation of laboratory evaluations in HIV vaccine trials. *Microbes Infect*, **7**, 1424–1432.

Goulder PJ, Phillips RE, Colbert RA, McAdam S, Ogg G, Nowak MA, Giangrande P, Luzzi G, Morgan B, Edwards A, McMichael AJ and Rowland-Jones S (1997). Late escape from an immunodominant cytotoxic T-lymphocyte response associated with progression to AIDS. *Nat Med*, **3**, 212–217.

Graham BS (2002). Clinical trials of HIV vaccines. *Annu Rev Med*, **53**, 207–221.

Gray RH, Wawer MJ, Brookmeyer R, Sewankambo NK, Serwadda D, Wabwire-Mangen F, Lutalo T, Li X, VanCott T and Quinn TC (2001). Probability of HIV-1 transmission per coital act in monogamous, heterosexual, HIV-1-discordant couples in Rakai, Uganda. *Lancet*, **357**, 1149–1153.

Hammon WM, Coriel LL and Wehrle PF (1983). Evaluation of Red Cross gamma globulin as a prophylactic agent for poliomyelitis. *JAMA*, **151**, 1272–1285.

Harari A, Petitpierre S, Vallelian F and Pantaleo G (2004). Skewed representation of functionally distinct populations of virus-specific CD4 T cells in HIV-1-infected subjects with progressive disease: changes after antiretroviral therapy. *Blood*, **103**, 966–972.

Hobeika AC, Morse MA, Osada T, Ghanayem M, Niedzwiecki D, Barrier R, Lyerly HK and Clay TM (2005). Enumerating antigen-specific T-cell responses in peripheral blood: a comparison of peptide MHC Tetramer, ELISpot, and intracellular cytokine analysis. *J Immunother*, **28**, 63–72.

Holl V, Peressin M, Decoville T, Schmidt S, Zolla-Pazner S, Aubertin AM and Moog C (2006). Nonneutralizing antibodies are able to inhibit human immunodeficiency virus type 1 replication in macrophages and immature dendritic cells. *J Virol*, **80**, 6177–6181.

Jagodzinski LL, Cooley JD, Weber M and Michael NL (2003). Performance characteristics of human immunodeficiency virus type 1 (HIV-1) genotyping systems in sequence-based analysis of subtypes other than HIV-1 subtype B. *J Clin Microbiol*, **41**, 998–1003.

Janetzki S, Cox J, Oden N and Ferrari G (2005a). Standardization and validation issues of the ELISPOT assay. In: A Kalyuzhny (ed), *Handbook of ELISPOT Methods and Protocols*, Minneapolis, MN, Humana Press, pp 51–86.

Janetzki S, Cox JH, Oden N and Ferrari G (2005b). Standardization and validation issues of the ELISPOT assay. *Methods Molec Biol*, **302**, 51–86.

Janeway C and Travers P (1997). Immunobiology: The Immune System in Health and Disease, New York, Garland Publishing.

Janssen EM, Lemmens EE, Wolfe T, Christen U, von Herrath MG and Schoenberger SP (2003). CD4+ T cells are required for secondary expansion and memory in CD8+ T lymphocytes. *Nature*, **421**, 852–856.

Jin X, Bauer DE, Tuttleton SE, Lewin S, Gettie A, Blanchard J, Irwin CE, Safrit JT, Mittler J, Weinberger L, Kostrikis LG, Zhang L, Perelson AS and Ho DD (1999). Dramatic rise in plasma viremia after CD8(+) T cell depletion in simian immunodeficiency virus-infected macaques. *J Exp Med*, **189**, 991–998.

Kaslow RA, Carrington M, Apple R, Park L, Munoz A, Saah AJ, Goedert JJ, Winkler C, O'Brien SJ, Rinaldo C, Detels R, Blattner W, Phair J, Erlich H and Mann DL (1996). Influence of combinations of human major histocompatibility complex genes on the course of HIV-1 infection. *Nat Med*, **2**, 405–411.

Kaul R, Dong T, Plummer FA, Kimani J, Rostron T, Kiama P, Njagi E, Irungu E, Farah B, Oyugi J, Chakraborty R, MacDonald KS, Bwayo JJ, McMichael A and Rowland-Jones SL (2001). CD8(+) lymphocytes respond to different HIV epitopes in seronegative and infected subjects. *J Clin Invest*, **107**, 1303–1310.

Kebba A, Kaleebu P, Rowland S, Ingram R, Whitworth J, Imami N and Gotch F (2004). Distinct patterns of peripheral HIV-1-specific interferon- gamma responses in exposed HIV-1-seronegative individuals. *J Infect Dis*, **189**, 1705–1713.

Khurana S, Needham J, Mathieson B, Rodriguez-Chavez IR, Catanzaro AT, Bailer RT, Kim J, Polonis V, Cooper DA, Guerin J, Peterson ML, Gurwith M, Nguyen N, Graham BS and Golding H (2006). Human immunodeficiency virus (HIV) vaccine trials: a novel assay for differential diagnosis of HIV infections in the face of vaccine-generated antibodies. *J Virol*, **80**, 2092–2099.

Kiepiela P, Smith AN and Rosenberg E (2005). Laboratory markers associated with progression of HIV infection. *Best Pract Res Clin Obstet Gynaecol*, **19**, 243–254.

Klausner RD, Fauci AS, Corey L, Nabel GJ, Gayle H, Berkley S, Haynes BF, Baltimore D, Collins C, Douglas RG, Esparza J, Francis DP, Ganguly NK, Gerberding JL, Johnston MI, Kazatchkine MD, McMichael AJ, Makgoba MW, Pantaleo G, Piot P, Shao Y, Tramont E, Varmus H and Wasserheit JN (2003). The need for a global HIV vaccine enterprise. *Science*, **300**, 2036–2039.

Koup RA, Safrit JT, Cao Y, Andrews CA, McLeod G, Borkowsky W, Farthing C and Ho DD (1994). Temporal association of cellular immune responses with the initial control of viremia in primary human immunodeficiency virus type 1 syndrome. *J Virol*, **68**, 4650–4655.

Letvin NL (2005). Progress toward an HIV vaccine. *Annu Rev Med*, **56**, 213–223.

Letvin NL, Mascola JR, Sun Y, Gorgone DA, Buzby AP, Xu L, Yang ZY, Chakrabarti B, Rao SS, Schmitz JE, Montefiori DC, Barker BR, Bookstein FL and Nabel GJ (2006). Preserved CD4+ central memory T cells and survival in vaccinated SIV-challenged monkeys. *Science*, **312**, 1530–1533.

Li F, Malhotra U, Gilbert PB, Hawkins NR, Duerr AC, McElrath JM, Corey L and Self SG (in press a). Peptide selection for human immunodeficiency virus type 1 CTL-based vaccine evaluation. *Vaccine*.

Li M, Gao F, Mascola JR, Stamatatos L, Polonis VR, Koutsoukos M, Voss G, Goepfert P, Gilbert P, Greene KM, Bilska M, Kothe DL, Salazar-Gonzalez JF, Wei X, Decker JM, Hahn BH and Montefiori DC (2005). Human immunodeficiency virus type 1 env clones from acute and early subtype B infections for standardized assessments of vaccine-elicited neutralizing antibodies. *J Virol*, **79**, 10108–10125.

Li M, Salazar-Gonzalez JF, Derdeyn CA, Morris L, Williamson C, Robinson JE, Decker JM, Li Y, Salazar MG, Polonis V, Mlisana K, Karim SA, Hong K, Greene KM, Bilska

M, Zhou J, Allen S, Chomba E, Mulenga J, Vwalika C, Gao F, Zhang M, Korber B, Hunter E, Hahn BH and Montefiori DC (in press b). Neutralization properties of acute and early Subtype C Human Immunodeficiency Virus Type 1 molecular env clones from heterosexually acquired infections in Southern Africa. *J Virol*.

Li Y, Svehla K, Mathy NL, Voss G, Mascola JR and Wyatt R (2006). Characterization of antibody responses elicited by human immunodeficiency virus type 1 primary isolate trimeric and monomeric envelope glycoproteins in selected adjuvants. *J Virol*, **80**, 1414–1426.

Lichterfeld M, Yu XG, Waring MT, Mui SK, Johnston MN, Cohen D, Addo MM, Zaunders J, Alter G, Pae E, Strick D, Allen TM, Rosenberg ES, Walker BD and Altfeld M (2004). HIV-1-specific cytotoxicity is preferentially mediated by a subset of CD8(+) T cells producing both interferon-gamma and tumor necrosis factor-alpha. *Blood*, **104**, 487–494.

Lopalco L, Barassi C, Paolucci C, Breda D, Brunelli D, Nguyen M, Nouhin J, Luong TT, Truong LX, Clerici M, Calori G, Lazzarin A, Pancino G and Burastero SE (2005). Predictive value of anti-cell and anti-human immunodeficiency virus (HIV) humoral responses in HIV-1-exposed seronegative cohorts of European and Asian origin. *J Gen Virol*, **86**, 339–348.

Lyons AB and Parish CR (1994). Determination of lymphocyte division by flow cytometry. *J Immunol Methods*, **171**, 131–137.

Maecker HT, Dunn HS, Suni MA, Khatamzas E, Pitcher CJ, Bunde T, Persaud N, Trigona W, Fu TM, Sinclair E, Bredt BM, McCune JM, Maino VC, Kern F and Picker LJ (2001). Use of overlapping peptide mixtures as antigens for cytokine flow cytometry. *J Immunol Methods*, **255**, 27–40.

Maecker HT, Rinfret A, D'Souza P, Darden J, Roig E, Landry C, Hayes P, Birungi J, Anzala O, Garcia M, Harari A, Frank I, Baydo R, Baker M, Holbrook J, Ottinger J, Lamoreaux L, Epling CL, Sinclair E, Suni MA, Punt K, Calarota S, El-Bahi S, Alter G, Maila H, Kuta E, Cox J, Gray C, Altfeld M, Nougarede N, Boyer J, Tussey L, Tobery T, Bredt B, Roederer M, Koup R, Maino VC, Weinhold K, Pantaleo G, Gilmour J, Horton H and Sekaly RP (2005). Standardization of cytokine flow cytometry assays. *BMC Immunol*, **6**, 13.

Mascola J (1999). Neutralization of HIV-1 infection of Peripheral Blood mononuclear Cells (PBMC). In: N Michael and JH Kim (eds), *HIV Protocols*, Minneapolis, MN, Humana Press, pp 317–322.

Mascola JR, Stiegler G, VanCott TC, Katinger H, Carpenter CB, Hanson CE, Beary H, Hayes D, Frankel SS, Birx DL and Lewis MG (2000). Protection of macaques against vaginal transmission of a pathogenic HIV-1/SIV chimeric virus by passive infusion of neutralizing antibodies. *Nat Med*, **6**, 207–210.

Mascola JR, Louder MK, Winter C, Prabhakara R, De Rosa SC, Douek DC, Hill BJ, Gabuzda D and Roederer M (2002). Human immunodeficiency virus type 1 neutralization measured by flow cytometric quantitation of single-round infection of primary human T cells. *J Virol*, **76**, 4810–4821.

Mascola JR, D'Souza P, Gilbert P, Hahn BH, Haigwood NL, Morris L, Petropoulos CJ, Polonis VR, Sarzotti-Kelsoe M and Montefiori DC (2005). Recommendations for the design and use of standard virus panels to assess neutralizing antibody responses elicited by candidate human immunodeficiency virus type 1 vaccines. *J Virol*, **79**, 10103–10107.

Mattapallil JJ, Douek DC, Buckler-White A, Montefiori D, Letvin NL, Nabel GJ and Roederer M (2006). Vaccination preserves CD4 memory T cells during acute simian immunodeficiency virus challenge. *J Exp Med*, **203**, 1533–1541.

McMichael AJ (2006). HIV vaccines. *Annu Rev Immunol*, **24**, 227–255.

McMichael AJ and Hanke T (2003). HIV vaccines 1983–2003. *Nat Med*, **9**, 874–880.

Mellors JW, Munoz A, Giorgi JV, Margolick JB, Tassoni CJ, Gupta P, Kingsley LA, Todd JA, Saah AJ, Detels R, Phair JP and Rinaldo CR, Jr (1997). Plasma viral load and CD4+ lymphocytes as prognostic markers of HIV-1 infection. *Ann Int Med*, **126**, 946–954.

Migueles SA, Sabbaghian MS, Shupert WL, Bettinotti MP, Marincola FM, Martino L, Hallahan CW, Selig SM, Schwartz D, Sullivan J and Connors M (2000). HLA B*5701 is highly associated with restriction of virus replication in a subgroup of HIV-infected long term nonprogressors. *Proc Natl Acad Sci USA*, **97**, 2709–2714.

Mindel A and Tenant-Flowers M (2001). ABC of AIDS: Natural history and management of early HIV infection. *Br Med J*, **322**, 1290–1293.

Mwau M, McMichael AJ and Hanke T (2002). Design and validation of an enzyme-linked immunospot assay for use in clinical trials of candidate HIV vaccines. *AIDS Res Hum Retroviruses*, **18**, 611–618.

Nishimura Y, Igarashi T, Haigwood N, Sadjadpour R, Plishka RJ, Buckler-White A, Shibata R and Martin MA (2002). Determination of a statistically valid neutralization titer in plasma that confers protection against simian–human immunodeficiency virus challenge following passive transfer of high-titered neutralizing antibodies. *J Virol*, **76**, 2123–2130.

Nishimura Y, Igarashi T, Haigwood NL, Sadjadpour R, Donau OK, Buckler C, Plishka RJ, Buckler-White A and Martin MA (2003). Transfer of neutralizing IgG to macaques 6 h but not 24 h after SHIV infection confers sterilizing protection: implications for HIV-1 vaccine development. *Proc Natl Acad Sci USA*, **100**, 15131–15136.

Pal R, Kalyanaraman VS, Nair BC, Whitney S, Keen T, Hocker L, Hudacik L, Rose N, Mboudjeka I, Shen S, Wu-Chou TH, Montefiori D, Mascola J, Markham P and Lu S (2006). Immunization of rhesus macaques with a polyvalent DNA prime/protein boost human immunodeficiency virus type 1 vaccine elicits protective antibody response against simian human immunodeficiency virus of R5 phenotype. *Virology*, **348**, 341–353.

Pantaleo G and Harari A (2006). Functional signatures in antiviral T-cell immunity for monitoring virus-associated diseases. *Nat Rev Immunol*, **6**, 417–423.

Pantaleo G and Koup RA (2004). Correlates of immune protection in HIV-1 infection: what we know, what we don't know, what we should know. *Nat Med*, **10**, 806–810.

Pantophlet R and Burton DR (2006). GP120: target for neutralizing HIV-1 antibodies. *Annu Rev Immunol*, **24**, 739–769.

Parren PW, Marx PA, Hessell AJ, Luckay A, Harouse J, Cheng-Mayer C, Moore JP and Burton DR (2001). Antibody protects macaques against vaginal challenge with a pathogenic R5 simian/human immunodeficiency virus at serum levels giving complete neutralization in vitro. *J Virol*, **75**, 8340–8347.

Perfetto SP, Chattopadhyay PK and Roederer M (2004). Seventeen-colour flow cytometry: unravelling the immune system. *Nat Rev Immunol*, **4**, 648–655.

Picker LJ and Maino VC (2000). The CD4(+) T cell response to HIV-1. *Curr Opin Immunol*, **12**, 381–386.

Pilcher CD, Fiscus SA, Nguyen TQ, Foust E, Wolf L, Williams D, Ashby R, O'Dowd JO, McPherson JT, Stalzer B, Hightow L, Miller WC, Eron JJ, Jr, Cohen MS and Leone PA (2005). Detection of acute infections during HIV testing in North Carolina. *N Engl J Med*, **352**, 1873–1883.

Pombo M, Berthold I, Gingrich E, Jaramillo M, Leef M, Sirota L, Hsu H and Arciniega J (2004). Validation of an anti-PA-ELISA for the potency testing of anthrax vaccine in mice. *Biologicals*, **32**, 157–163.

Quinn TC, Wawer MJ, Sewankambo N, Serwadda D, Li C, Wabwire-Mangen F, Meehan MO, Lutalo T and Gray RH (2000). Viral load and heterosexual transmission of human immunodeficiency virus type 1. Rakai Project Study Group. *N Engl J Med*, **342**, 921–929.

Roederer M, Brenchley JM, Betts MR and De Rosa SC (2004). Flow cytometric analysis of vaccine responses: how many colors are enough? *Clin Immunol*, **110**, 199–205.

Rosenberg ES, Billingsley JM, Caliendo AM, Boswell SL, Sax PE, Kalams SA and Walker BD (1997). Vigorous HIV-1-specific CD4+ T cell responses associated with control of viremia. *Science*, **278**, 1447–1450.

Russell ND, Hudgens MG, Ha R, Havenar-Daughton C and McElrath MJ (2003). Moving to human immunodeficiency virus type 1 vaccine efficacy trials: defining T cell responses as potential correlates of immunity. *J Infect Dis*, **187**, 226–242.

Samri A, Durier C, Urrutia A, Sanchez I, Gahery-Segard H, Imbart S, Sinet M, Tartour E, Aboulker JP, Autran B and Venet A (2006). Evaluation of the interlaboratory concordance in quantification of human immunodeficiency virus-specific T cells with a gamma interferon enzyme-linked immunospot assay. *Clin Vaccine Immunol*, **13**, 684–697.

Scherer A, Frater J, Oxenius A, Agudelo J, Price DA, Gunthard HF, Barnardo M, Perrin L, Hirschel B, Phillips RE and McLean AR (2004). Quantifiable cytotoxic T lymphocyte responses and HLA-related risk of progression to AIDS. *Proc Natl Acad Sci USA*, **101**, 12266–12270.

Schmidt G, Amiraian K, Frey H, Stevens RW and Berns DS (1987). Densitometric analysis of Western blot (immunoblot) assays for human immunodeficiency virus antibodies and correlation with clinical status. *J Clin Microbiol*, **25**, 1993–1998.

Schmitz JE, Kuroda MJ, Santra S, Sasseville VG, Simon MA, Lifton MA, Racz P, Tenner-Racz K, Dalesandro M, Scallon BJ, Ghrayeb J, Forman MA, Montefiori DC, Rieber EP, Letvin NL and Reimann KA (1999). Control of viremia in simian immunodeficiency virus infection by CD8+ lymphocytes. *Science*, **283**, 857–860.

Shapiro HM (2003). *Practical Flow Cytometry*. New York, John Wiley and Sons, Inc.

Shedlock DJ and Shen H (2003). Requirement for CD4 T cell help in generating functional CD8 T cell memory. *Science*, **300**, 337–339.

Spearman P (2003). HIV vaccine development: lessons from the past and promise for the future. *Curr HIV Res*, **1**, 101–120.

Sun JC and Bevan MJ (2003). Defective CD8 T cell memory following acute infection without CD4 T cell help. *Science*, **300**, 339–342.

Sun Y, Iglesias E, Samri A, Kamkamidze G, Decoville T, Carcelain G and Autran B (2003). A systematic comparison of methods to measure HIV-1 specific CD8 T cells. *J Immunol Methods*, **272**, 23–34.

Suss J and Sinnecker H (1991). Immune reactions against rabies viruses – infection and vaccination. *Exp Pathol*, **42**, 1–9.

Travers K and Kanki PJ (1990). HIV antibody detection in serum. In: A Aldovini and BD Walker (eds), *Techniques in HIV Research*, New York, Stockton Press, pp 6–10.

Trigona WL, Clair JH, Persaud N, Punt K, Bachinsky M, Sadasivan-Nair U, Dubey S, Tussey L, Fu TM and Shiver J (2003). Intracellular staining for HIV-specific IFN-gamma production: statistical analyses establish reproducibility and criteria for distinguishing positive responses. *J Interferon Cytokine Res*, **23**, 369–377.

Trkola A, Kuster H, Rusert P, Joos B, Fischer M, Leemann C, Manrique A, Huber M, Rehr M, Oxenius A, Weber R, Stiegler G, Vcelar B, Katinger H, Aceto L and Gunthard HF (2005). Delay of HIV-1 rebound after cessation of antiretroviral therapy through passive transfer of human neutralizing antibodies. *Nat Med*, **11**, 615–622.

Wang S, Pal R, Mascola JR, Chou TH, Mboudjeka I, Shen S, Liu Q, Whitney S, Keen T, Nair BC, Kalyanaraman VS, Markham P and Lu S (2006). Polyvalent HIV-1 Env vaccine formulations delivered by the DNA priming plus protein boosting approach are effective in generating neutralizing antibodies against primary human immunodeficiency virus type 1 isolates from subtypes A, B, C, D and E. *Virology*, **350**, 34–47.

Wilson NA, Reed J, Napoe GS, Piaskowski S, Szymanski A, Furlott J, Gonzalez EJ, Yant LJ, Maness NJ, May GE, Soma T, Reynolds MR, Rakasz E, Rudersdorf R, McDermott AB, O'Connor DH, Friedrich TC, Allison DB, Patki A, Picker LJ, Burton DR, Lin J, Huang L, Patel D, Heindecker G, Fan J, Citron M, Horton M, Wang F, Liang X, Shiver JW, Casimiro DR and Watkins DI (2006). Vaccine-induced cellular immune responses reduce plasma viral concentrations after repeated low-dose challenge with pathogenic simian immunodeficiency virus SIVmac239. *J Virol*, **80**, 5875–5885.

Zhu T, Corey L, Hwangbo Y, Lee JM, Learn GH, Mullins JI and McElrath MJ (2003). Persistence of extraordinarily low levels of genetically homogeneous human immunodeficiency virus type 1 in exposed seronegative individuals. *J Virol*, **77**, 6108–6116.

15 The future direction of cell-based assays

Uma Prabhakar and Marian Kelley

Clinical Pharmacology and Experimental Medicine, 145 King of Prussia Road, Radnor, PA 19087, USA

Since cells are a more accurate representation of the physiological state *in vivo*, offer the possibility of dynamic experimentation, and are cost-effective, cell-based assays serve as an early biological filter in various stages of discovery and development (Diaz-Mochon et al, 2007). A number and variety of cell-based assays are well established that measure cell proliferation, toxicity, motility, production of a measurable product, and morphology. The criteria for conducting "GLP-like" methods for a large number of these applications have been described in much detail in previous chapters of this book.

The combined efforts in the areas of combinatorial chemistry and genomics have significantly increased the number of compounds and therapeutic targets/candidates available for screening in cell-based assays. It is expected that these numbers will likely reach into the million range in the near future and provide vast chemical diversity for drug discovery. This reservoir of chemical diversity creates obvious downstream hurdles for any screening effort. Additionally, with reduced animal experimentation, cell-based assays will assume an even more prominent role in the drug development process calling for increasing demand for higher-quality output, well-characterized and well-maintained cells, and more consistent and precise analytical data. Some examples exemplifying these applications are discussed in this chapter and the challenges of applying stringent laboratory practices are addressed.

The existing traditional high throughput screening (HTS) techniques, such as the use of 96-well microtiter plates, will have to be upgraded to meet the increased demands of screening. Miniaturization of cell-based assays, which allow for greater throughput, while concurrently reducing cost, is already the future wave and has been successfully applied to bacterial and soluble protein base assays. A recent publication (Angres, 2005) summarizes the future of the cell-based microarrays and reviews their various applications.

Moving ahead, the focus of screening methods is expected to be on easy-to-use and highly sensitive assays that provide continuous records of cellular activity. Although, most research activities currently concentrate on drug discovery, cell-based assays are expected to gain popularity in diagnostics, molecular biology, biochemistry and neuroscience, genetics, toxicology studies, bioengineering, and proteomics, among other fields as they become more cost effective and prove to be biologically significant.

As we move to automated screening assays, cells will have to be treated as one of the reagents, albeit a very delicate reagent that is sensitive to environmental conditions, evaporation, liquid handling and more. Choosing the optimum cell system early in assay development based on the biology of the target and the nuances of the screen is a critical, although not always a straightforward task. As with living organisms, cells may change their biological properties, resulting in misleading data based on any experimental variability, such as temperature, pH, media or serum concentration. Some cells may experience diminished functional capacity over time. Therefore, maintaining a documented cell bank, knowing the characteristics of the cell lines in advance and understanding how the protein is regulated in continuous culture is essential for cell-based assays.

The area of cell-based microarrays for the identification and profiling of cell membrane composition and properties is receiving much attention lately. This application is being used to identify and quantitate carbohydrate-mediated cellular adhesion to the use of peptide–major histocompatibility complex (peptide–MHC) microarrays that allow the detection of specific T cells that recognize disease-related events (Stone et al, 2005). Identification of T cell epitopes is a vital but often slow and difficult step in studying the immune response to infectious agents and autoantigens. This system uses microarrays of immobilized, recombinant MHC–peptide complexes, co-stimulatory molecules, and cytokine-capture antibodies. The array elements act as synthetic antigen-presenting cells and specifically elicit T cell responses, including adhesion, secretion of cytokines, and modulation of surface markers. The method allows facile identification of pertinent T cell epitopes in a large number of candidates and simultaneous determination of the functional outcome of the interaction and has been applied to characterizing the activation of human CD4(+) and CD8(+) T cells responding to vaccinia, influenza, HIV-1, and Epstein–Barr viruses.

The recently failed clinical trial of a recombinantly expressed humanized superagonist anti-CD28 mAb, TGN14122 (Suntharalingam et al, 2006) has resulted in the need for wider investigations that are relevant to the safety evaluations of such agents before they can be considered safe for use in humans. One such investigation entails the use of a novel *in vitro* procedure where human white blood cells are presented with the antibody drug to simulate the striking release of cytokines and profound lymphocyte proliferation activation that mimics a "cytokine storm" and may predict the toxicity of such drug agents (Stebbings et al, 2007). These examples offer just a sampling of the types of cell-based assays that will have to be developed and validated to address emerging questions from drug discovery and development.

It is obvious that some of the cell-based assays currently being developed to address specific questions on safety, immune function and related to drug action and/or efficacy, are extremely complicated. The development and validation of methodologies to measure specific responses using cells are equally challenging. Depending upon the stage of drug development, biological relevance and, how close the specific response measured is to being a biomarker or a surrogate or a diagnostic, the assay validation will vary in its rigor.

Overall, biomarker analyses in general are exploratory in nature, diverse and varied in their applications, and are not based on any 'official' guidelines for validation of laboratory biomarker assays. This is in sharp contrast to bioanalytical method validations that support pharmacokinetic (PK) assessments of conventional small molecule drugs (Guidance for industry, Bioanalytical method validation, 2001) or macromolecules (DeSilva et al, 2003). Human specimens for diagnosis, prevention, or treatment of any disease or impairment or for the assessment of the health of individual patients are certified under Clinical Laboratory Improvement Amendments (CLIA). The Clinical and Laboratory Standards Institute (CSLI) publishes standard practices for CLIA certification.

As a consequence, there are consistent adaptations of related regulations in bioanalytical and clinical laboratory procedures for use in cell-based assays or biomarker research. Recently, members of the American Association of Pharmaceutical Scientists (AAPS) Ligand Binding Assay Bioanalytical Focus Group Biomarker Subcommittee described a "fit-for-purpose" approach for biomarker method development and validation (Lee et al, 2006). The fit-for-purpose validation strategy is continuous and evolving and is adapted for the intended application. The rigor of method increases as biomarker data from such assays are used for increasingly advanced clinical or business-critical decision making. The focus of the fit-for-purpose validation strategy that is described in the Lee at al (2006) publication is on soluble biomarkers.

The analytical challenges encountered when adopting regulatory guidelines for biomarker/diagnostic or pharmacokinetic analysis is well known among investigators active in this area. Here we propose that cell-based assay method development and validation, particularly for the newer applications, also adopt a fit-for-purpose strategy. It seems redundant to evaluate extensive parameters for a validation irrespective of how the information will be used. When a method is being used to identify drug targets during screening, a certain level of rigor and robustness should be built into the method. On the other hand, if a cell-based biomarker is being evaluated during an early stage of drug development on a purely exploratory basis, the rigor may be different and must be defined. Thus, it makes the most sense that assay validation be tailored to meet the intended purpose of the biomarker study with a level of rigor commensurate with the intended use of that data.

Each method must be considered within the context of how and where the data will be used. The validity of the data being generated must stand up to some level of scrutiny, whether it is used to support early discovery screening or the drug label. As novel cell-based methods are introduced to aid in the development of drugs, care must be taken to ensure that the data generated are accurate and consistent. A "GLP-like" environment dictates that the method of conduct will be well thought out and well documented, both facets of good science, independent of the stage of drug development. In order to address cell-based assay development specifically, it will be most efficient to convene a group of investigators engaged in the development of cell-based assays to propose a fit-for-purpose strategy as they apply to cells and tissues and yet adequately meet regulatory requirements.

References

Angres B (2005). Cell microarrays. *Expert Rev Mol Diagn*, **5,** 769–779.

DeSilva B, Smith W, Weiner R et al (2003). Recommendations for the bioanalytical method validation of ligand binding assays to support pharmacetuic assessments of macromolecules. *Pharm Res*, **20,** 1885–1900.

Diaz-Mochon JJ, Tourniarire G and Bradley M (2007). Microarray platforms for enzymatic and cell-based assays. *Chem Soc Rev*, **36,** 449–457.

Guidance for industry, Bioanalytical method validation (2001). US department of Health and Human Services, Food & Drug Evaluation and Research (CDER), Central for Veterinary Medicine (CVM).

Lee JW, Devnarayan, V, Barrett YC, Weiner R, Allinson J, Fountain S, Keller S, Weinryb I, Green M, Duan L, Rogers JA, Miliham R, O'Brien PJ, Sailstad J,

Khan M, Ray C and Wagner JA (2006). Fit-for-purpose method development and validation for successful biomarker measurement. *Pharm Res*, 23(2): 312–327.

Stebbings R, Findlay L, Eastwood EC, Bird C, North D, Mistry Y, Dilger P, Liefooghe E, Cludts I, Fox B, Tarrant G, Robinson J, Meager T, Dolman C, Thorpe SJ, Bristow Am Wadhwa M, Thorpe R and Poole S (2007). "Cytokine storm" in the phase I trial of monoclonal antibody TGN1412: better understanding the causes to improve preclinical testing of immunotherapeutics. *J Immunol*, **179**, 3325–3331.

Stone JD, Demokowicz WE and Stern LJ (2005). HLA-restricted epitope identification and detection of functional T cell responses by using MHC–peptide and co-stimulatory microarrays. *Proc Natl Acad Sci USA*, **102**, 3744–3749.

Suntharalingam G, Perry MR, Ward S, Brett SJ, Castello-Cortes A, Brunner MD et al (2006). Cytokine storm in a phase 1 trial of the anti-CD28 monoclonal antibody TGN1412, *N Engl J Med* **355**, 1018–1028.

Index

Validation of Cell-Based Assays in the GLP Setting: A Practical Guide Editors, Uma Prabhakar and Marian Kelley